Research as Development Perspective

Research as Development Perspective

Editors

Victoria Samanidou
George Zachariadis

MDPI • Basel • Beijing • Wuhan • Barcelona • Belgrade • Manchester • Tokyo • Cluj • Tianjin

Editors
Victoria Samanidou
Aristotle University of Thessaloniki
Greece

George Zachariadis
Aristotle University
Greece

Editorial Office
MDPI
St. Alban-Anlage 66
4052 Basel, Switzerland

This is a reprint of articles from the Special Issue published online in the open access journal *Separations* (ISSN 2297-8739) (available at: https://www.mdpi.com/journal/separations/special_issues/confer_develop).

For citation purposes, cite each article independently as indicated on the article page online and as indicated below:

LastName, A.A.; LastName, B.B.; LastName, C.C. Article Title. *Journal Name* **Year**, *Article Number*, Page Range.

ISBN 978-3-03936-930-0 (Hbk)
ISBN 978-3-03936-931-7 (PDF)

Contents

About the Editors

Victoria Samanidou was born on the 11th of January 1963, in Thessaloniki, Greece. In 1985, she obtained her Bachelor of Science in Chemistry from Aristotle University of Thessaloniki, Greece, and in 1990 obtained a doctorate (PhD) in Chemistry from Aristotle University of Thessaloniki. The topic of her thesis was the distribution and mobilization of heavy metals in waters and sediments from rivers in Northern Greece. In the same year, Dr. Samanidou joined the Laboratory of Analytical Chemistry as a Technical Assistant in the Department of Chemistry at Aristotle University of Thessaloniki. Nine years later, she was elected as Lecturer in the Laboratory of Analytical Chemistry and in 2007 she joined the Institute of Analytical Chemistry and Radiochemistry in Graz Technical University for four months to develop methods using LC-MS/MS. Since 2015, Dr. Samanidou has been a Full Professor in the Laboratory of Analytical Chemistry in the Department of Chemistry at Aristotle University of Thessaloniki, where she currently serves as Director of the Laboratory. Further, she has authored and co-authored more than 170 original research articles in peer-reviewed journals, and 45 reviews and 50 chapters in scientific books, with H-index 36 (Scopus June 2020, Author ID: 7003896015) and ca 3500 citations. She has supervised four PhD Theses, 24 postgraduate Diploma Theses, 2 postdoc researchers, and more than 15 undergraduate Diploma Theses. She has served as a member on 10 advisory PhD committees, 21 examination PhD committees, and 32 examination committees for postgraduate Diploma Theses. She is an editorial board member for more than 10 scientific journals and she has reviewed ca 500 manuscripts in more than 100 scientific journals. She was also a guest editor in more than 10 Special Issues for various scientific journals. She served as the Academic Editor for *Separations* (MDPI), a Regional Editor in *Current Analytical Chemistry*, and as Editor in Chief of *Pharmaceutica Analytica Acta*. Her research interests include the development and validation of analytical methods for the determination of inorganic and organic substances using chromatographic techniques; the development and optimization of methodology for sample preparation of various samples, e.g., food, biological fluids, etc.; the study of new chromatographic materials used in separation and sample preparation (e.g., polymeric sorbents, monoliths, carbon nanotubes, fused core particles, etc.) when compared to conventional materials. She has also organized scientific committee for over 20 scientific conferences. In December 2015, Dr. Samanidou was elected as President of the Steering Committee of the Division of Central and Western Macedonia of the Greek Chemists' Association. In November 2018, she was reelected to serve the same leading position for three more years. A milestone in her career came in 2016, when she was included in top 50 power list of women in *Analytical Science*, as proposed by Texere Publishers.

George Zachariadis is a Professor in the Department of Chemistry of Aristotle University of Thessaloniki. For three years, he served as the Director of the Laboratory of Analytical Chemistry for the last five years has been the head of the postgraduate program in Chemistry, where he teaches quantitative chemical analysis, instrumental chemical analysis, chemometrics metrology and quality control, and archaeometry and bioanalysis. Zachariadis also teaches analytical chemistry courses to students in the Department of Pharmacy. Moreover, he is involved in teaching and tutoring in two graduate degree programs in Chemical Analysis, Quality Control, and Bioanalysis. He has supervised 6 doctoral dissertations, 20 postgraduate dissertations, and more than 40 senior theses. He is the author or co-author of 13 teaching books. He has also published an international book entitled

Inductively Coupled Plasma Atomic Emission Spectrometry. He is the author or co-author of about 150 announcements presented at international and national conferences and 140 original research and review papers published in international scientific journals in the fields of chemical, bioanalytical, pharmaceutical, food, archaeometric, and environmental analysis. There are currently more than 3300 citations of his scientific work. He acts as a reviewer for about 30 international journals. He has served as a Guest Editor in a Special Issue for the journal *Current Analytical Chemistry*. His research includes the development of hyphenated analytical techniques, various separation techniques, atomic spectrometry and mass spectrometry, speciation in metal analysis and analyte determination in biological substrates, as well as environmental, technological, and archaeological materials. Regarding chemometrics, he developed implementation methods of various statistical tools to optimize analytical methods and interpret scientific results and data. He has participated in or supervised about 35 research projects and cooperated with several national and international research institutes and laboratories.

Preface to "Research as Development Perspective"

This Special Issue, "Research as a Development Perspective", is dedicated to data presented at the first Conference in Chemistry for Graduate/Postgraduate Students and PhD candidates at Aristotle University of Thessaloniki, which was the outcome of research conducted by young chemists in Northern Greece. The conference was organized by the Chemistry Department at Aristotle University of Thessaloniki, the Association of Greek Chemists-Division of Central and Western Macedonia, and the Association of Chemists in Northern Greece. The scope of this conference was to provide young chemists (but also last year's students) with the opportunity to be well prepared for their next career steps in an increasingly demanding job market. Moreover, they had the possibility of presenting their scientific results to a large audience, which strengthened their soft skills. Lastly, the active engagement of students in the organization of the conference enhanced their teamwork abilities, a highly valuable when developing professional maturity.

Scientific Topics covered in Special Issue

- Analytical Chemistry-Quality Control
- Archaeometry-Maintenance and restoration of cultural monuments
- Physical-Theoretical and Computational Chemistry
- Toxicology- Medicinal Chemistry
- Environmental Chemistry and Technology-Pollution Control
- Food Analysis
- Chemical Technology-Materials- Green Chemistry

The Guest Editors wish to express their gratitude to Separations–MDPI for sponsoring the publication of articles presented in the Conference. They also wish to thank all authors for their fine contribution and, last but not least, the organizers: the Chemistry Department-Aristotle University of Thessaloniki, the Association of Greek Chemists-Division of Central and Western Macedonia, and the Association of Chemists in Northern Greece.

Victoria Samanidou, George Zachariadis
Editors

Article

Non-Destructive X-ray Spectrometric and Chromatographic Analysis of Metal Containers and Their Contents, from Ancient Macedonia

Christos S. Katsifas [1,2,*], Despina Ignatiadou [3], Anastasia Zacharopoulou [1], Nikolaos Kantiranis [4], Ioannis Karapanagiotis [5] and George A. Zachariadis [2]

1 Laboratory of Physico-Chemical Studies & Archaeometry, Archaeological Museum of Thessaloniki, 54013 Thessaloniki, Greece; anastasiazach@hotmail.com
2 Laboratory of Analytical Chemistry, Department of Chemistry, Aristotle University, 54124 Thessaloniki, Greece; zacharia@chem.auth.gr
3 Department of Sculpture, National Archaeological Museum, 10682 Athens, Greece; dignatiadou@culture.gr
4 Department of Mineralogy, Petrology and Economic Geology, School of Geology, Aristotle University, 54124 Thessaloniki, Greece; kantira@geo.auth.gr
5 Department of Management and Conservation of Ecclesiastical Cultural Heritage Objects, University Ecclesiastical Academy of Thessaloniki, N. Plastira 65, 54250 Thessaloniki, Greece; y.karapanagiotis@aeath.gr
* Correspondence: chkatsifas@culture.gr, Tel.: +30-6944-262-282

Received: 1 April 2018; Accepted: 31 May 2018; Published: 11 June 2018

Abstract: This work describes a holistic archaeometric approach to ancient Macedonian specimens. In the region of the ancient city Lete, the deceased members of a rich and important family were interred in a cluster of seven tombs (4th century BC). Among the numerous grave goods, there was also a set of metal containers preserving their original content. The physico-chemical analysis of the containers and their contents was performed in order to understand the purpose of their use. For the containers, Energy Dispersive micro-X-Ray Fluorescence (EDμXRF) spectroscopy was implemented taking advantage of its non-invasive character. The case (B35) and the small pyxis (B37) were made of a binary Cu-Sn alloy accompanied by a slight amount of impurities (Fe, Pb, As) and the two miniature bowls were made of almost pure Cu. For the study of the contents, a combination of EDμXRF, X-Ray Diffraction (XRD), and Gas Chromatography—Mass Spectrometry (GC-MS) was carried out. Especially for the extraction of the volatile compounds, the Solid Phase Micro-Extraction (SPME) technique was used in the headspace mode. Because of the detection of Br, High Pressure Liquid Chromatography coupled to a Diode-Array-Detector (HPLC-DAD) was implemented, confirming the existence of the ancient dye shellfish purple (porphyra in Greek). The analytical results of the combined implementation of spectrometric and chromatographic analytical techniques of the metal containers and their contents expand our knowledge about the pharmaceutical practices in Macedonia during the 4th century BC.

Keywords: Derveni; Ancient Macedonia; micro-XRF; XRD; HPLC-DAD; HS-SPME/GC-MS; ancient medicines; ancient pharmaceuticals; shellfish purple; porphyra; high-tin bronzes; bronzes

---

1. Introduction

The Derveni tombs were accidentally revealed in 1962, 9.5 km NW of Thessaloniki, Macedonia, Greece. The six cist graves and one Macedonian tomb that were excavated then had not been looted and contained rich offerings, mainly dated to the 4th century BC. The deceased were members of a rich and important Thessalian family that probably lived in the nearby ancient city Lete. In grave A, the so-called Derveni papyrus was found; fragments of a papyrus roll with the most important Orphic religious text of the 4th century BC, preserving excerpts of an earlier poem. The biggest and richest grave was grave B. It contained the cremated remains of a man and his female consort, which had been placed in an elaborate bronze vessel, today known as the famous Derveni krater. That male individual was an important member of the elite, probably a royal companion who died when he was approximately 35–50 years old. In addition to the bronze krater, the burial contained a gold wreath and other gold jewelry, twenty silver vessels, many bronze vessels, stone alabastra, glass vessels and pottery vases, the iron weapons of the dead, a folding board gaming set with glass gaming counters, pieces of a leather cuirass, bronze greaves, and a gold coin of King Philip II [1,2].

Among the numerous grave goods, from grave B, was a lidded box (B35), preserving its original content (Figure 1). It is a semi-cylindrical case divided into three compartments. Each compartment is filled with a mass of "clay". The case has a hinged lid that protects the contents. According to the first estimations [1], B35 is a case for storing cosmetics and its contents are materials for makeup. Ongoing research on the history of medicine in Macedonia and the comparison with other metal cases which have been unearthed in Macedonian burials [3,4], after B35, had led to suspicions that the Derveni case was a medical case. Other metal finds from the same grave that were more or less associated with the case B35 are: two bowls (B43a, B43b) and a pyxis (B37), all preserving their original content. In the two bowls, there are preserved pieces of a thin dark-coloured cake and in the pyxis there is a red powder.

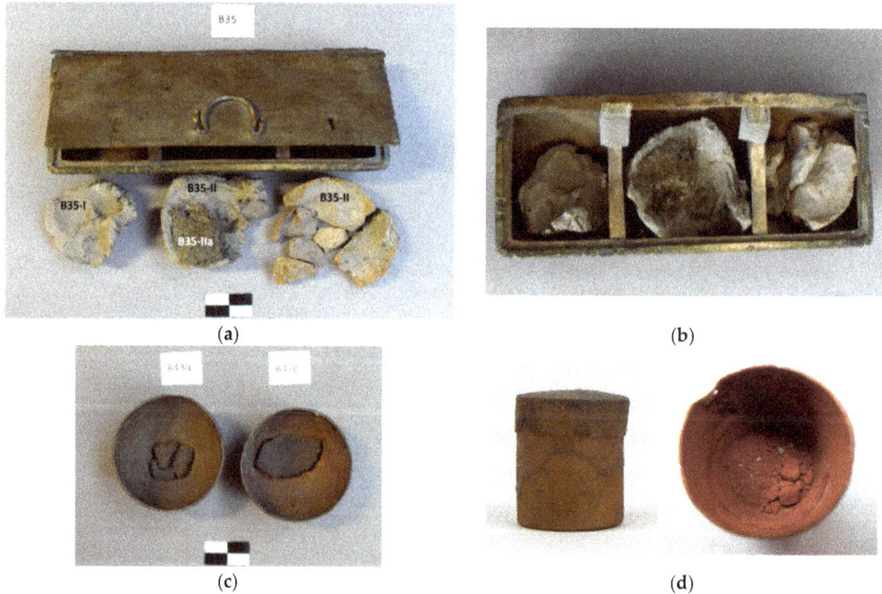

Figure 1. Metal containers and their contents. (**a**) Lidded case B35 and its content (cakes B35-I, B35-II, B35-III). (**b**) Case B35 without the lid. (**c**) Bowls B43a and B43b. (**d**) Lidded pyxis B37 (with and without lid).

The aims of the present physico-chemical study are: (a) to determine the composition of the alloy used for the construction of the metal containers, as well as the differentiation according to each part of the artefact (e.g., body, lid, handle, nails); and (b) to identify the inorganic and organic components of the contents using chemical and mineralogical analysis. The analytical results will contribute to the effort to determine the nature of these artefacts and consequently their purpose of use, as well as to illuminate the identity of their owner.

2. Materials and Methods

The metal containers under study are: one lidded case (B35), two miniature bowls (B43a, B43b), and a small pyxis (B37), all found in Derveni grave B. The metal case B35 consists of different parts: body, lid, handle, the lid's hinges, the rim of the case proper, and the nails the latter is fastened with. The bowls B43a and B43b have been made from a single metal sheet and pyxis B37 consists of its body and a lid. Case B35 is preserved in very good condition. It has been chemically cleaned in the past and does not carry any corrosion products. Its color, after the cleaning, is golden. On the other hand the two bowls (B43a, B43b) are reddish, denoting a differentiation of the alloy compared to that of B35. They are also preserved in a very good state and apparently do not present corrosion layers. Finally, pyxis B37 presents, at the body, a thick reddish corrosion layer in combination with extensive restoration works. Its lid maintains a better condition with areas where the metal does not present extensive corrosion layers.

For the analysis of the metal containers, the choice of a non-invasive and non-destructive analytical technique is obligatory. Sampling is largely prohibited, according to the Greek Archaeological Law, due to the uniqueness, integrity, and small size of the artefacts. So the choice of a non-invasive technique like Energy Dispersive micro X-ray Fluorescence (EDμXRF) spectroscopy—which is a widespread technique for the analysis of ancient metals [5,6]—was a requisite instead of a technique that provides bulk analysis, such as Inductively Coupled Plasma Spectroscopy (ICP) or Atomic Absorption Spectroscopy (AAS). On the other hand, EDμXRF provides a chemical profile of the surface, which may differ from the bulk composition and perhaps is not representative of the whole [7,8]. When bronze artefacts are exposed to the atmosphere or are buried in the ground, their surface acquires a more or less thick patina under which the metal core may remain substantially unchanged [9]. Before the implementation of the EDμXRF analysis, optical macro- and microscopic examination of each artefact was applied in order to select the spots for analysis. Areas free of corrosion products or materials from conservation treatments were selected. Especially at the pyxis B37, mechanical removal of the corrosion products took place in order to measure from the bulk of the metal.

The contents of case B35 bear similar hues. Starting from left to right, they were numbered: B35-I, B35-II, and B35-III. Especially at the middle compartment of B35 and on the top of cake B35-II, a fourth cake can be seen which has a different hue than the others. Instead of an earthy hue, it is reddish and was hence numbered B35-IIa. Macroscopically, cakes B35-I, B35-II, and B35-III resemble dried out clays. Respectively, the contents of bowls B43a and B43b were numbered B43a-I and B43b-I. These cakes are harder than those in case B35 and their hue is brownish red with a dark top. To characterize the components of the four cakes that are preserved in the lidded case B35, as well as the two from the bronze bowls (B43a, B43b) and the red powder from the pyxis (B37), a physico-chemical analysis was undertaken in combination with a mineralogical examination. After the preliminary morphological examination by stereomicroscopy, EDμXRF spectroscopy was carried out for the analysis of the inorganic constituents. In order to determine their mineralogical composition, XRD diffractometry was implemented. For the study of the organic constituents, the samples were extracted with the Head Space—Solid Phase Micro-Extraction (HS-SPME) technique and the absorbed volatiles were analyzed by Gas Chromatography—Mass Spectrometry (GC-MS). Because of the significance of the material under study, an effort has been made to implement the above analytical techniques, as much as possible, in a non-destructive way. Analysis using High Pressure Liquid Chromatography coupled to a Diode-Array-Detector (HPLC-DAD) was carried out in only one sample from cake B35 IIa in order to ascertain the constituent which is responsible for its color. The cause of this implementation was the interesting results of EDμXRF in combination with its reddish hue.

2.1. Energy Dispersive Micro-X-ray Fluorescence spectrometry (EDμXRF)

The instrument ARTAX 400 (Bruker AXS Microanalysis GmbH, Berlin, Germany) was used for the implementation of the micro-X-ray Fluorescence spectroscopy technique (μXRF). This Energy Dispersive spectrometer has been specially designed for the demands of archaeometry [10].

The measuring head, which is placed on a x,y,z—motor driven positioning stage, is comprised of: (a) an air cooled Mo X-ray fine focus tube, (b) a peltier cooled silicon drift detector (SDD), and (c) a Charge Coupled Device (CCD) camera for the visual inspection of the sample. The X-ray beam is restricted by a collimator and on the surface of the sample has a diameter between 200 μm and 1500 μm, depending on the type of collimator used. All excitation and detection paths can be rinsed with Helium (He) gas which is directed, by use of two gas jets, towards the sample surface in order to reduce the absorption of the beam through the air and thus improve the excitation conditions for elements with a lower atomic number. Acquired spectra were processed with software SPECTRA 7.4. This program enables the acquisition of measurement data including the control of all Artax 400 components. Parallel to the measurement, it is possible to carry out qualitative elemental analysis, data reduction (integral calculation of the different spectral lines through deconvolution), and calculation of the artefact elemental concentration.

For the analysis of the metal containers, EDμXRF measurements (ARTAX 400, Bruker AXS Microanalysis GmbH, Berlin, Germany) were taken in order to determine their elemental composition. For the metal case B35, measurements were taken from every different part (body, lid, hinge, rim, nails). Each final reported result is the mean value of three measurements. The measurement time for each spot analysis was 100s. The voltage of the exciting X-ray beam was 50 kV and the current was 700 μA. The collimator with a 1500 μm diameter was used in order to get a representative result and to avoid alterations due to micro structural inhomogeneity [11,12]. Especially for the analysis of the pyxis B37, which presents corrosion layers, and in order to avoid the removal of corrosion in an extended area (since it is considered as a destructive action), the collimator with a diameter of 650 μm was used. The certified data from Certified Reference Material (CRM) 32XSN1 (MBH—Analytical Ltd., Barnet, UK) was used as a calibration file. The quantified results were balanced to 100%. In order to check the accuracy of the implemented method, measurements were taken with the same parameters as those employed for the artefacts, from Certified Reference Materials (CRMs). BCR-691 (European Commission—Joint Research Centre, Institute for Reference Materials and Measurements, Brussels, Belgium), a set of five (5) copper alloys, was used (Table 1).

Table 1. Results of the analysis of Certified Reference Materials (CRM's) and detection limits (wt %).

BCR-691		**Sn**	**Zn**	**Pb**	**As**
Quaternary bronze (A)	Certified value ± unsertainty	7.16 ± 0.21	6.02 ± 0.22	7.90 ± 0.7	0.19 ± 0.01
	Measured value ± std	7.00 ± 0.3	6.20 ± 0.45	7.50 ± 0.50	0.20 ± 0.00
Brass (B)	Certified value ± unsertainty	2.06 ± 0.07	14.80 ± 0.50	0.39 ± 0.04	0.10 ± 0.01
	Measured value ± std	2.00 ± 0.10	14.50 ± 0.21	0.30 ± 0.06	0.10 ± 0.00
Arsenic copper (C)	Certified value ± unsertainty	0.202 ± 0.029	0.05 ± 0.005	0.175 ± 0.014	4.6 ± 0.27
	Measured value ± std	0.17 ± 0.06	0.03 ± 0.01	0.18 ± 0.04	4.5 ± 0.50
Lead bronze (D)	Certified value ± unsertainty	10.1 ± 0.80	0.148 ± 0.024	9.2 ± 1.7	0.285 ± 0.022
	Measured value ± std	9.60 ± 0.50	0.17 ± 0.02	9.05 ± 0.416	0.26 ± 0.05
Tin bronze (E)	Certified value ± unsertainty	7 ± 0.60	0.157 ± 0.025	0.204 ± 0.018	0.194 ± 0.02
	Measured value ± std	7.33 ± 0.50	0.17 ± 0.01	0.2 ± 0.00	0.2 ± 0.00
Estimated detection limits (%)		0.02	0.01	0.03	0.01

For the elemental analysis of the contents of the metal containers, with the EDμXRF technique, the voltage of the exciting X-ray beam was 35 kV and the current was 900 μA. It was also implemented in an He gas atmosphere. The 1500 μm diameter collimator was used to counterbalance the inhomogeneity of the materials under study. Each final reported result is the mean value of seven measurements since the contents present greater inhomogeneity than the metal containers. For the quantification of the analytical results, the CRM soil sample SO-3 (Canada Centre for Mineral & Energy Technology, Ottawa, ON, Canada) was used in combination with software SPECTRA 7.4 (Bruker AXS Microanalysis GmbH, Berlin, Germany).

2.2. X-ray Diffraction (XRD)

The content of the metal containers were archaeological material and therefore were studied in their bulk form without any grinding, homogenization, or pretreatment. In order to identify their mineralogical composition, X-ray diffraction analysis (XRD) was applied directly on the surface of the samples.

A Phillips PW1820/00 diffractometer equipped with a PW1710/00 microprocessor (PHILIPS, Almelo, The Netherlands) was used and the samples were scanned over the 3°–63° 2θ interval at a scanning speed of 1.2°/min. Semi-quantitative analysis estimates of the abundance of the mineral phases were derived from the XRD data, using the intensity of a certain reflection [13], the density, and the mass absorption coefficient for Cu-K_α radiation for the minerals present. The identification of the minerals present was made using the DDView$^+$/Sileve$^+$ ICDD's viewing/search indexing software provided with the PDF-4$^+$ (PDF-4$^+$, 2009, Powder diffraction file^{TM}, International Centre for Diffraction Data, Newtown Square, PA, USA) relational database. Corrections were made using the external standard mixtures of minerals (i.e., standard mixtures of the identified minerals in the studied samples such as quartz and cristobalite, feldspars, micas and chlorite, amphibole, calcite and sulfate minerals, as well as iron oxides). The detection limit of the method was ±1% *w*/*w* [14]. The degree of crystallinity and the calculation of the amorphous phase amount were calculated according to the method described by Kantiranis et al. (2004) [14].

2.3. Head Space—Solid Phase Micro-Extraction/Gas Chromatography—Mass Spectrometry (HS—SPME/GC—MS)

The pre-treatment and extraction of the samples was done by the HS-SPME technique using a polydimethylsiloxane (PDMS) coated fiber (100 μm film) in the head space above the heated samples. GC-MS (Agilent 6873 K gas chromatograph—Agilent 5973 quadrupole mass detector, Agilent Technologies, Santa Clara, CA, USA) analysis of the absorbed volatiles from the samples was carried out and finally the identification of the compounds was succeeded by using the NIST library. The GC-MS operating conditions are listed in Table 2.

Table 2. GC-MS operating conditions.

Operating Conditions	Value
GC model	Agilent 6873 K gas chromatograph (Electron Ionization mode)
Column DB-5MS (capillary)	30 m × 0.25 mm × 0.10 μm
Injector port temperature	200 °C (SPME fiber was remained there for 6 min)
Carrier gas, Flow-rate	Helium, 2.0 mL/min (constant pressure at 29.8 psi)
Oven temperature program	60 °C (3 min) to 270 °C, 10 °C/min ramp time
Transfer line temperature	250 °C
MS model	Agilent 5973 quadrupole mass detector (Scan mode)
Total time of chromatographic analysis	40 min

2.4. High Pressure Liquid Chromatography—Diode Array Detector (HPLC-DAD)

High Pressure Liquid Chromatography was coupled to a Diode-Array-Detector (Ultimate 3000, Dionex, Sunnyvale, CA, USA,) and consisted of a LPG-3000 quaternary HPLC pump with a vacuum degasser, a WPS-3000SL auto sampler, a column compartment TCC-3000SD, and a UV-Vis Diode Array Detector (DAD-3000) (Dionex, Sunnyvale, CA, USA) Analyses were carried out by injecting 20 μL into an Alltima (Grace-Alltech, Deerfield, IL, USA) HP C18 (250 mm × 3 mm, i.d. 5 μm) column at a stable temperature of 35 °C.

Two solvent reservoirs, containing (A) water + 0.1% (*v*/*v*) Tri-Fluoro-Acetic Acid (TFA) and (B) acetonitrile + 0.1% (*v*/*v*) TFA, were used under a gradient elution program which was developed and evaluated for the analysis of shellfish purple components offering extremely low limits of detection [15].

The sample was immersed in a hot (80 °C) dimethyl sulfoxide (DMSO) bath and kept there for 15 min. After centrifugation, the upper liquid phase was immediately submitted to HPLC.

By employing this method, polar and apolar compounds are detected. The method was previously devised and optimised for the extraction and solubilisation of shellfish purple, as described in detail elsewhere [16].

3. Results

3.1. Metal Containers

According to the results of Table 3, the main parts (body, lid, and rim) of the metal case B35 were made of copper-tin (Cu-Sn) alloy. Lead (Pb), iron (Fe), and calcium (Ca) were detected, at all parts, at quantities much lower than 1 wt %. Arsenic (As) is also detected, except at the four nails which present significant differentiation, at their chemical composition, compared to the main parts of B35. They consist of almost pure Cu which is detected at levels higher than 99.0 wt %. Nickel (Ni) and titanium (Ti) are only detected at the nails of B35, at very low quantities (rounded average for the four nails are: 0.17 ± 0.03 and 0.01 ± 0.00 wt % for Ni and Ti, respectively), while Sn is present at levels below 1 wt %. Another part of B35 that presents a significant differentiation is the hinge of the lid where the quantity of Sn (6.56 ± 0.30 wt %) is very low compared to the body of the case (12.11 ± 0.20 wt %).

Bowls B43a and B43b present the same qualitative and quantitative analysis. Cu is detected at levels around 99 wt %. The two bowls present also the same chemical elements Sn, Pb, Fe, As, Ni, and Ca at similar quantities (lower than 0.5 wt %). At pyxis B37, measurements were only taken from its lid after the removal of superficial corrosion products. The body presents thick corrosion layers which will not provide us with a safe quantitative result. The lid of pyxis B37 mainly consists of Cu (86.96 ± 0.40 wt %) and Sn (12.16 ± 0.20 wt %) at quantities similar to the main parts of the metal case B35, e.g., the body (Cu: 87.61 ± 0.60 wt % and Sn: 12.16 ± 0.20 wt %). Similar to B35, Pb and Fe are the minor elements, at quantities lower than 1 wt %. However, Ni is only detected at a very low quantity (0.15 ± 0.50 wt %).

3.2. Contents of the Metal Containers

According to the EDμXRF analysis (Table 4), cakes B35-I, B35-II, and B35-III, which present a similar appearance, texture, and colour, also present similar analytical results. Major constituents (concentration higher than 1 wt %) are: SiO_2, Al_2O_3, $Fe_2O_{3(total)}$, MgO, CaO, CuO, and K_2O; and minor constituents (concentration between 0.1 wt % and 1.0 wt %) are: V_2O_3, and MnO. Finally, at the level of traces (concentration lower than 0.1 wt %), the following were detected: Cr_2O_3, NiO, ZnO, Rb_2O, SrO, and PbO.

Cake B35-IIa, which presents a reddish hue, presents some differentiations compared to the other cakes. Al_2O_3, which is a main constituent in the other cakes, is a minor one at B35-IIa (0.28 ± 0.30 wt %). Furthermore, the total percentage of the inorganic constituents is lower in than the other cakes and ranges from 16.56 wt % to 42.27 wt %. Finally, B35-IIa is the only cake where the chemical element bromine (Br) was detected (Figure 2). This result, in combination with the differentiation of the colour of the cake (reddish hue), created suspicions for the possible existence of a dye. In the cakes (B43a-I, B43b-I) of the two bowls, there is a similar distribution between major and minor constituents and those that detected as traces. Compared to cakes of the metal case B35, the total percentage of the inorganic oxides (30.98 wt % for cake B43a-Iand 15.03 wt % for cakeB43b-I) is significantly lower. The powder in pyxis B37 presents great differentiations, in terms of the analytical results, compared to the contents of the others containers. There are only two major constituents: $Fe_2O_{3(t)}$ (11.44 ± 1.10 wt %) and CuO (9.71 ± 1.10 wt %). The minor constituent detected is CaO, and ZnO and K_2O are only detected in traces.

Table 3. Chemical composition (wt % ± std) of metal containers—results of EDµXRF analysis (* nd: not detected).

No	Description	Cu	Sn	Pb	Fe	As	Ni	Ca	Ti
1	B35-body	87.61 ± 0.60	12.11 ± 0.20	0.19 ± 0.04	0.06 ± 0.02	0.02 ± 0.01	nd*	0.01 ± 0.00	nd
2	B35-lid	87.68 ± 0.50	12.07 ± 0.30	0.12 ± 0.02	0.08 ± 0.03	0.04 ± 0.01	nd	0.02 ± 0.01	nd
3	B35-lid-handle	85.33 ± 0.30	14.42 ± 0.20	0.15 ± 0.03	0.06 ± 0.02	0.01 ± 0.00	nd	0.03 ± 0.01	nd
4	B35-lid-hinge	93.00 ± 0.60	6.59 ± 0.30	0.23 ± 0.04	0.13 ± 0.05	0.03 ± 0.01	nd	0.02 ± 0.01	nd
5	B35 body-rim	83.44 ± 0.30	15.62 ± 0.50	0.72 ± 0.04	0.04 ± 0.02	0.01 ± 0.00	nd	0.02 ± 0.01	nd
6	B35-nail-01	99.20 ± 0.60	0.06 ± 0.01	0.35 ± 0.03	0.20 ± 0.03	nd	0.17 ± 0.03	0.01 ± 0.00	0.01 ± 0.00
7	B35-nail-02	98.95 ± 0.50	0.76 ± 0.20	0.40 ± 0.02	0.18 ± 0.04	nd	0.17 ± 0.03	0.02 ± 0.00	0.01 ± 0.00
8	B35-nail-03	99.22 ± 0.50	0.10 ± 0.02	0.34 ± 0.04	0.15 ± 0.02	nd	0.16 ± 0.03	0.02 ± 0.01	nd
9	B35-nail-04	99.09 ± 0.60	0.12 ± 0.40	0.43 ± 0.06	0.17 ± 0.08	nd	0.16 ± 0.03	0.01 ± 0.00	0.01 ± 0.00
10	B43a-body	99.20 ± 0.40	0.04 ± 0.01	0.45 ± 0.05	0.14 ± 0.05	0.01 ± 0.00	0.16 ± 0.03	0.01 ± 0.00	nd
11	B43b-body	99.12 ± 0.50	0.04 ± 0.01	0.48 ± 0.02	0.18 ± 0.03	0.01 ± 0.00	0.17 ± 0.03	0.01 ± 0.00	nd
12	B37-lid	86.96 ± 0.40	12.16 ± 0.20	0.66 ± 0.04	0.07 ± 0.02	nd	0.15 ± 0.05	nd	nd

Table 4. Contents of metal containers—results of the EDµXRF analysis (wt %).

Description	SiO_2	TiO_2	Al_2O_3	$Fe_2O_{3(t)}$	V_2O_5	Cr_2O_3	MnO	MgO	CaO	CuO	NiO	ZnO	Rb_2O	SrO	PbO	K_2O	Br	TOTAL
B35-I	42.66 ± 2.10	1.42 ± 0.30	15.53 ± 1.20	10.38 ± 1.10	0.18 ± 0.06	0.09 ± 0.04	0.31 ± 0.10	2.19 ± 0.20	4.90 ± 0.42	2.30 ± 0.32	0.01 ± 0.00	0.01 ± 0.00	0.02 ± 0.01	0.04 ± 0.02	0.01 ± 0.00	3.36 ± 0.45	nd	**83.41**
B35-II	29.84 ± 1.40	0.82 ± 0.20	10.75 ± 0.40	5.95 ± 0.40	0.10 ± 0.04	0.04 ± 0.02	0.18 ± 0.02	0.86 ± 0.40	3.12 ± 0.25	4.12 ± 0.70	nd	0.01 ± 0.00	0.01 ± 0.00	0.02 ± 0.01	0.02 ± 0.01	1.84 ± 0.28	nd	**57.70**
B35-IIa	16.58 ± 1.20	0.43 ± 0.12	0.28 ± 0.30	4.65 ± 1.00	0.04 ± 0.03	0.15 ± 0.05	0.06 ± 0.03	0.90 ± 0.20	4.67 ± 0.32	11.85 ± 1.56	0.03 ± 0.01	0.02 ± 0.01	0.02 ± 0.01	0.01 ± 0.00	0.01 ± 0.00	1.02 ± 0.04	0.40 ± 0.05	**41.14**
B35-III	43.94 ± 2.15	1.22 ± 0.30	18.06 ± 1.80	9.92 ± 1.10	0.22 ± 0.04	0.07 ± 0.03	0.10 ± 0.02	1.84 ± 0.35	2.00 ± 0.24	2.02 ± 0.35	0.01 ± 0.00	0.01 ± 0.00	0.02 ± 0.01	0.02 ± 0.01	0.14 ± 0.03	2.92 ± 0.67	nd	**82.53**
B43a-I	6.48 ± 1.20	0.35 ± 0.15	2.19 ± 0.25	3.57 ± 0.70	0.75 ± 0.25	0.12 ± 0.02	0.03 ± 0.01	0.86 ± 0.25	3.65 ± 0.18	12.02 ± 1.32	0.04 ± 0.02	0.01 ± 0.00	nd	0.01 ± 0.00	0.16 ± 0.00	0.73 ± 0.25	nd	**30.98**
B43b-I	4.43 ± 0.80	0.22 ± 0.30	1.10 ± 0.20	2.42 ± 0.40	0.44 ± 0.10	0.07 ± 0.03	0.01 ± 0.00	0.63 ± 0.18	1.78 ± 0.16	3.39 ± 0.88	0.03 ± 0.01	nd	nd	0.01 ± 0.00	0.10 ± 0.00	0.41 ± 0.25	nd	**15.03**
B37-I	nd	nd	nd	11.44 ± 1.10	nd	nd	nd	nd	0.27 ± 0.08	9.71 ± 1.10	nd	0.02 ± 0.00	nd	nd	nd	0.02 ± 0.00	nd	**21.47**

nd: not detected.

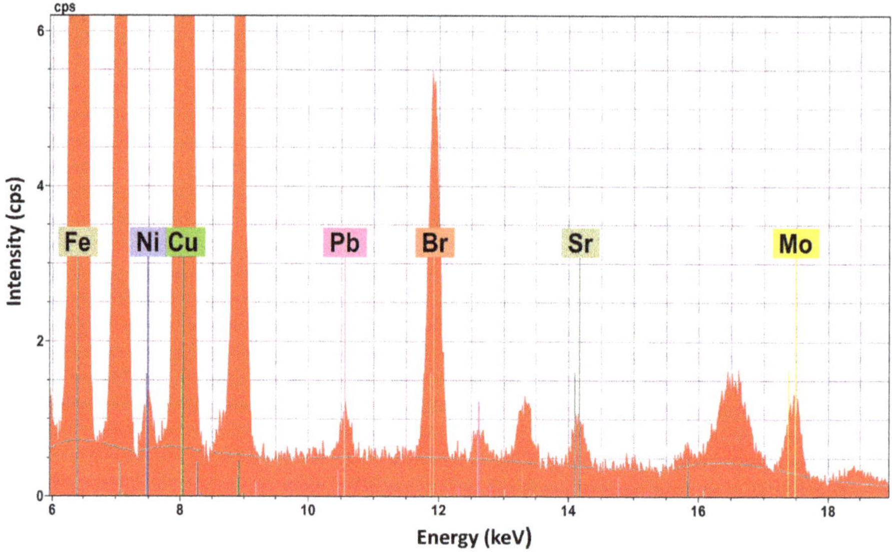

Figure 2. EDμXRF spectrum of cake B-35-IIa—detection of Br (peak K_{a1} at 11.88 keV and K_{b1} at 13.29 keV).

The detection of Br in combination with the colour differentiation of the sample B35-IIa led to the investigation of the possible existence of a dye that contains Br. The implementation of the HPLC-DAD technique resulted in the detection of the compound 6,6′-dibromoindigotin (DBI) (Figure 3), which was detected at 288 nm. The limit of daltons is 2.4×10^{14}, as described in detail elsewhere [13].

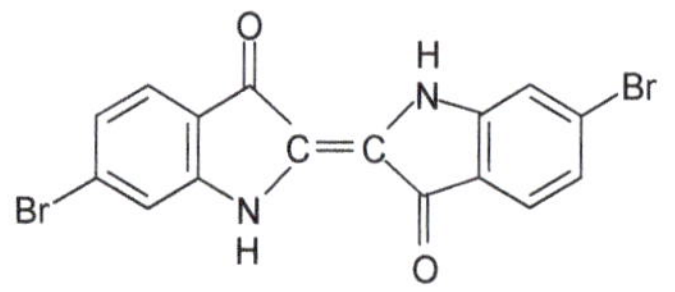

Figure 3. 6,6′-dibromoindigotin (DBI).

According to the results of the mineralogical analysis (Table 5) of the cakes in case B35, similar amounts of quartz (SiO_2) (39–44 wt %) were detected at high quantities. Other constituents in abundance are: plagioclase (8–29 wt %), mica (6–14 wt %) that belongs to the group of silicate minerals, and chlorite (4–8 wt %). Minerals like K-feldspar and calcite are only detected in cakes B35-II and B35-III. Amorphous matter estimated at high level ranges from 16% to 31%, except in cake B35-I, where it is only 5 wt %. In cakes B43a-I and B43b-I, the amorphous matter estimated at even higher levels ranges from 65 wt % to 78 wt %. Except for the siliceous minerals (quartz, plagioclase, and mica), cristobalite (6–12 wt %), gypsum (5 wt %), and graphite (1 wt %) were also detected. Graphite was possible to identify (Figure 4) and measure using the 89-8487 ICDD card (main peak 3.3540Å, in comparison with the 3.3434 Å main peak of quartz, 46-1045 ICDD card). Especially 2 wt % bassanite was detected in cake B43b-I, a calcium sulfate mineral. Finally, powder B37-I presented a very different mineralogical composition than the others. The dominant constituents are the ferrous oxides (hematite and magnetite: 71 wt %), a copper sulphate mineral antlerite (15 wt %), gypsum (8 wt %), bassanite (5 wt %), and a low quantity of a rare borate mineral: inderite (1 wt %).

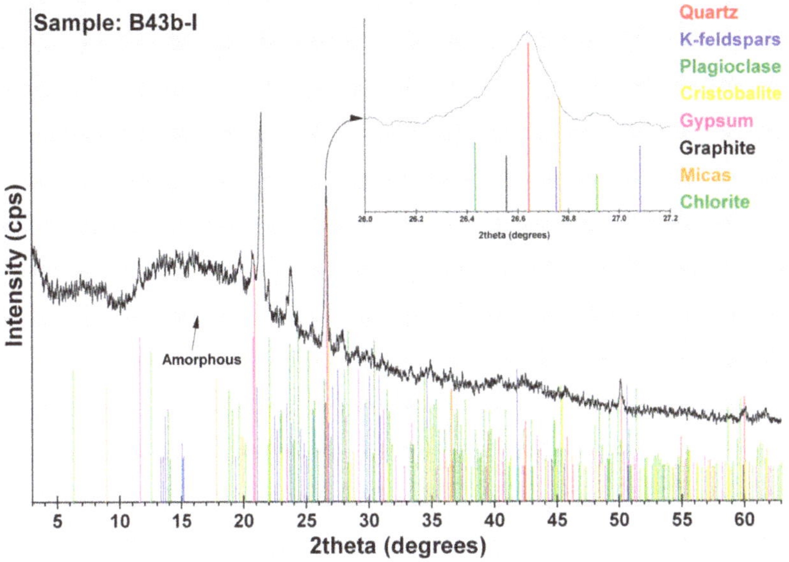

Figure 4. XRD pattern of cake B43b-I.

Table 5 presents the organic constituents of the contents of the metal containers according to the results of the HS-SPME/GC-MS analysis. In cake B35-I, only three (3) organic compounds were detected, none were detected at B35-II, and only three (3) (most of them fatty acids) were detected at B35-III. Contrary to the above, in cake B35-IIa, which presents a reddish hue, twenty-two (22) organic constituents (fatty acids, fatty acid esters, hydrocarbons, quinolines, phthalate esters, and amines) were detected (Figure 5). At contents B43a-I and B43b-I, the same three organic constituents were detected and at the red powder B37-I, seven (7) (thioamides, lactam, hydrocarbons, halide, and phthalate ester) were detected. Except for B43a-I and B43b-I, the other contents do not share any of the organic compounds. Fatty acids are the class of organic compounds which is common in the B35 and B43 contents. On the other hand, classes like thioamides and lactam are only detected at powder B37.

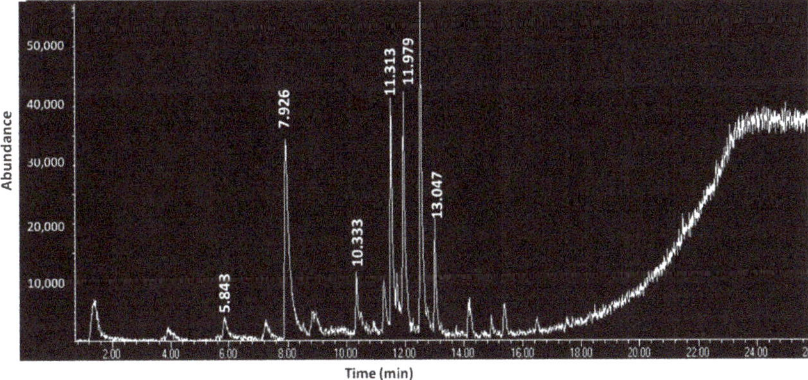

Figure 5. GC-MS chromatogram of volatile constituents of the sample B35-IIa (retention times correspond to analytes as described in Table 6).

Table 5. Contents of metal containers—results of the XRD analysis (wt %).

Sample	Quartz	Plagioclase	Mica	Chlorite	Dolomite	Amorphous	K-Feldspar	Calcite	Amphibole	Cristobalite	Gypsum	Graphite	Hydrozincite	Hematite	Magnetite	Bassanite	Inderite	Antlerite
B35-I	39	28	14	8	nd	5	nd	nd	5	nd	nd	nd	1	nd	nd	nd	nd	nd
B35-II	43	8	9	5	nd	18	7	7	3	nd	nd	nd	nd	nd	nd	nd	nd	nd
B35-IIa	42	9	8	7	3	31	nd	nd	nd	nd	nd	nd	nd	nd	nd	nd	nd	nd
B35-III	44	13	6	4	nd	16	8	6	3	nd	nd	nd	nd	nd	nd	nd	nd	nd
B43a-I	7	1	1	1	nd	78	nd	nd	nd	6	5	1	nd	nd	nd	nd	nd	nd
B43b-I	11	1	1	nd	nd	65	2	nd	nd	12	5	1	nd	nd	nd	2	nd	nd
B37-I	nd	nd	nd	nd	nd	nd	nd	nd	nd	nd	8	nd	nd	62	9	5	1	15

nd: not detected

Table 6. Contents of the metal containers: results of HS-SPME/GC-MS analysis.

No	RT (min)	Chemical Compound	Chemical Formula	B35-I	B35-II	B35-IIa	B35-III	B43a-I	B43b-I	B37-I
1	1.205	Ethanethioamide	C_2H_5NS	nd	nd	nd	nd	nd	nd	+
2	1.422	2,4-Hexadiyne	C_6H_6	nd	nd	+	nd	nd	nd	nd
3	1.827	2,4,6,8-tetramethyl-1-undecene	$C_{15}H_{30}$	nd	nd	nd	nd	nd	nd	+
4	1.841	Caprolactam	$C_6H_{11}NO$	nd	nd	nd	nd	nd	nd	+
5	2.587	Pentane, 2,3-dimethyl	C_7H_{16}	nd	nd	nd	nd	nd	nd	+
6	3.136	Napthalene-1-isocyano	$C_{11}H_7N$	nd	nd	nd	nd	nd	nd	+
7	3.151	1-Octanol, 2-butyl-	$C_{12}H_{26}O$	nd	nd	+	nd	nd	nd	nd
8	3.939	Propanoic acid, 2-methyl-,butylester	$C_8H_{16}O_2$	nd	nd	+	nd	nd	nd	nd
9	5.848	Pentanoic acid	$C_{19}H_{30}O_3$	nd	nd	+	nd	nd	nd	nd
10	7.267	Didodecyl phthalate	$C_{32}H_{54}O_4$	nd	nd	+	nd	nd	nd	nd
11	7.926	1,3-Dioxolane-2-propanoic acid, 2-methyl-,ethylester	$C_9H_{16}O_4$	nd	nd	+	nd	nd	nd	nd
12	8.855	Hexane, 3,3-dimethyl	C_8H_{18}	nd	nd	+	nd	nd	nd	nd
13	10.333	Nonadecane	$C_{19}H_{40}$	nd	nd	+	nd	nd	nd	nd
14	10.925	2,5,8,-Triphenylbenzotristriazole	$C_{24}H_{15}N_9$	nd	nd	+	nd	nd	nd	nd
15	10.932	8-Benzylquinoline	$C_{16}H_{13}N$	nd	nd	+	nd	nd	nd	nd
16	10.962	Benzimidazo[2,1-a]isoquinoline	$C_{15}H_{10}N_2$	nd	nd	+	nd	nd	nd	nd
17	11.105	Butane, 1-chloro-3,3-dimethyl	$C_6H_{13}Cl$	nd	nd	nd	nd	+	+	+
18	11.249	1,2-Benzenedicarboxylic acid, bis[2-methylpropyl] ester	$C_{16}H_{22}O_4$	nd	nd	+	nd	nd	nd	nd
19	11.284	Phthalic acid, 6-ethyl-3-octylisobutylester	$C_{22}H_{34}O_4$	nd	nd	+	nd	nd	nd	nd
20	11.313	1,2-Benzenedicarboxylic acid, bis[2-methylpropyl] ester	$C_{16}H_{22}O_4$	nd	nd	+	nd	nd	nd	nd

Table 6. *Cont.*

No	RT (min)	Chemical Compound	Chemical Formula	B35-I	B35-II	B35-IIa	B35-III	B43a-I	B43b-I	B37-I
21	11.569	6-Undecylamine	$C_{11}H_{25}N$	nd	nd	+	nd	nd	nd	nd
22	11.935	1,4-Oxathiin, 2,3-dihydro-6-methyl	C_5H_8OS	nd	nd	+	nd	nd	nd	nd
23	11.979	Formic acid, ethoxymethylene hydrazide	$C_4H_8N_2O_2$	nd	nd	+	nd	nd	nd	nd
24	12.509	Dibutylphthalate	$C_{16}H_{22}O_4$	nd	nd	nd	nd	+	+	+
25	12.515	Dibutylphthalate	$C_{16}H_{22}O_4$	nd	nd	+	nd	nd	nd	nd
26	12.520	1,4-Benzenedicarboxylic acid, bis[2-methylpropyl] ester	$C_{16}H_{22}O_4$	nd	nd	+	nd	nd	nd	nd
27	12.586	Dibutylphthalate	$C_{16}H_{22}O_4$	nd	nd	+	nd	nd	nd	nd
28	13.007	Sulfurous acid, butyldecylester	$C_{14}H_{30}O_3S$	nd	nd	nd	nd	+	+	nd
29	13.047	Nonadecane, 2-methyl-	$C_{20}H_{42}$	nd	nd	+	nd	nd	nd	nd
30	14.166	Morpholineethanamine	$C_6H_{14}N_2O$	nd	nd	+	nd	nd	nd	nd
31	14.203	Thiophene3-(1,1-dimethylethoxy)	$C_8H_{12}OS$	nd	nd	+	nd	nd	nd	nd
32	14.956	*p*-Terphenyl	$C_{18}H_{14}$	nd	nd	+	nd	nd	nd	nd
33	15.417	2-Piperidinecarboxylic acid, methylester, PFP	$C_{10}H_{12}F_5NO_3$	nd	nd	+	nd	nd	nd	nd
34	16.485	1-(Prop-2-ynyl)-3,3-bis(trifluoromethyl)diaziridine	$C_6H_4F_6N_2$	nd	nd	+	nd	nd	nd	nd
35	24.458	Boron, bis{μ-[(3,5-bis(1,1-dimethylethyl)-1 H-pyrazolato-N1 N2]}diethyl-	$C_{26}H_{50}B_2N_4$	nd	nd	nd	+	nd	nd	nd
36	25.072	10-Ethyl-3-chloro-7-(*N*-*p*-nitrobenzyldeneamino)-phenothiazine	$C_{21}H_{16}ClN_3O_2S$	nd	nd	nd	+	nd	nd	nd
37	27.215	Yohimban-16-carboxylic acid, 17-hydroxy-10-methoxy-, methylester ...	$C_{22}H_{28}N_2O_4$	+	nd	nd	nd	nd	nd	nd
38	28.978	Hexadecanoic acid, 2-hydroxy-1,3-propanediylester	$C_{35}H_{68}O_5$	+	nd	nd	nd	nd	nd	nd
39	30.384	9-Octadecanoic acid (Z)-2 hydroxy-1,3-propanediylester	$C_{39}H_{72}O_5$	nd	nd	nd	+	nd	nd	nd
40	38.123	9-Octadecanoic acid (Z)-2 hydroxy-3- propanediylester	$C_{37}H_{70}O_5$	+	nd	nd	nd	nd	nd	nd

+: detected, nd: not detected.

4. Discussion

4.1. Metal Containers

Metal case B35, which was found in Derveni grave B, presents differentiation in terms of the chemical composition of its various parts (Table 2). These parts have been made from hammered metal sheets (except the lid's handle which was cast). In general when Sn exceeds the limit of 1% in a Cu alloy, it is characterized as a deliberate addition [17]. The main parts of B35 (body, lid, the lid's handle, and the rim) were made of tin bronze (Cu-Sn) alloy accompanied by a low amount of impurities (Pb, Fe, As, Ca). This type of alloy was employed in Macedonia during the Classical period for the manufacture of vessels and utensils [7,18,19]. There is a classification of bronzes from low to high Sn depending on the amount of Sn that was added to the alloy. According to the related literature, there is no single definition of the concentration of Sn that would define an ancient bronze artefact as a "high-tin" [8,12,20]. The most accepted compromise, to characterize an ancient bronze as "high-tin", is an Sn concentration between 14 wt % and 16 wt %. [21,22]. Given that the lid's handle and the rim contain enough Sn (14.42–15.62 wt %) to characterize them as "high-tin" bronzes, the body and lid are close to this definition since they contain around 12 wt % Sn. On the contrary, the hinge contains only 6.59 ± 0.30 wt % Sn. The addition of Sn to Cu increases the strength and hardness of the alloy [23–25]. It also increases the resistance of the bronze artefact to corrosive parameters [21]. On the other hand, a Sn content higher than 15 wt % strongly increases the embrittlement of the alloy, raising problems (e.g., cracking) during hammering [18,25]. The role of the quantity of Sn is also significant to the appearance of bronze artefacts. A high Sn content gives them a "golden" hue which probably imitates vessels of precious metals [8,21,26] and denotes the high social status of the owner of B35; already suggested for grave goods in Derveni grave B [1]. Only the four nails are different and those contain Sn as an impurity and not as an additive [17]. Its very low level (0.06–0.76 wt %) makes the alloy easy to work with [27], but the authors also express their reservation as to whether the nails and hinge are ancient, or are modern additions. The chemical composition of bowls B43a and B43b, which are also hammered, is also different. Their main constituent is Cu (at level of 99 wt %) accompanied by a low amount of impurities (Sn, Pb, Fe, Ni, Ca). Sn is detected at very low quantities (0.04 ± 0.01 wt %) and is characterized as an impurity and not as an additive [17]. Finally, the lid of pyxisB37 presents a relatively high amount of Sn (12.16 ± 0.20 wt %), similar to the lid of case B35 (12.07 ± 0.01 wt %). The criteria of the ancient smith and their customers were more the social status and prestige imparted by the appearance of the objects [28] (ensuring the golden appearance of the artefacts with the addition of a relatively high amount of Sn) and not so much the workability. The smith of the 4th century BC probably developed skills to work with such brittle alloys without causing any cracks. The fact that pyxis B37 presents extensive corrosion compared to the good preservation of case B35 must be attributed to the taphonomic environment (e.g, the coexistence with other metal artefacts [27]) and not so with the chemical composition.

Fe is detected in all metal artefacts of the present study. In all cases, it is characterized as an impurity due to its very low concentration (0.04–0.20 wt %) [23,29]. Fe readily enters Cu during smelting and differences in Fe content arise through differences in the smelting process. The implementation of a simple and short smelting process results in an average Fe content of around 0.05 wt % instead of a process that involves slagging where the average Fe content rises at 0.5 wt % [30,31]. In this study, it is estimated that the first procedure was implemented. Moreover, for cold hammering procedures, Fe must not exceed 0.5 wt % because the formability of the artefact decreases due to precipitation phenomena [32]. Finally, the detection of very low Fe concentrations is an indicator that native and unrefined Cu was used [23,30].

Pb is considered as an impurity for all results of Table 2 since its concentration is lower than 2 wt % [33]. This is an expected result for hammered objects but not for cast ones (e.g., B35 lid's handle which contains only 0.15 ± 0.03 wt % Pb). This result is unexpected since Pb was a common additive in ancient bronzes. It improved fluidity and castability as it causes a reduction of the melting point. Most Hellenistic bronze cast artefacts contain more than 2 wt % Pb [34].

Arsenic was detected at all parts of B35 case (except the four nails) and bowls B43a and B43b at very low levels (0.01–0.04 wt %). This very small quantity may indicate that the initial ore deposit was not rich in As [33]. According to a previous study [7], Greek bronzes of the 4th century BC contain, in general, less than 1 wt % As, which is confirmed by the present investigation.

4.2. Contents of the Metal Containers

Elements like Si, Al, Fe Mg, Ca, and K which were detected as the main constituents, according to the EDμXRF analysis (Table 4), are in abundance in the earth's crust [21,27] and are consequently common soil elements. The only chemical element that is not common and is detected in all contents is Cu. A possible interpretation for this is contamination from the metal containers due to the long-term coexistence at the burial environment [35]. The other result, not common in soil, is the detection of Br in cake B35-IIa. Br was only detected in this cake, which also presents a reddish hue. The consequent implementation of the HPLC-DAD technique led to the detection of the compound 6,6′-dibromoindigotin (DBI), which is used as an index for the identification of shellfish purple (or porphyra in Greek or mollusc purple). Purple was the colour of royalty and a designator of social status.

According to the EDμXRF results of (Table 4), the powder of pyxis B37 presents a very different chemical composition. Elements which are the main constituents in the other contents (Si, Al, Mg, Ca, K) have not been detected in B37-I. Fe and Cu oxides comprises almost 20 wt % of the powder. The rest of it could be organic material, carbon oxides, or chemical elements of a very low atomic number (lower than 12) for which the excitation conditions of the EDμXRF technique are poor [36]. The high content of powder B37-I in ferrous constituents was also confirmed by the XRD results (Table 5), which have been normalized at 100 wt % contrary to the results of EDμXRF (Table 4). According to them, B37-I is mostly comprised of hematite (Fe_2O_3), magnetite (Fe_3O_4), and antlerite ($Cu_3(SO_4)(OH)_4$). Hematite is a red mineral pigment well known in antiquity [37]. Antlerite is a possible corrosion product of the reaction of the bronze antiquities, which are located in urban areas, with sulfur dioxide (a common pollutant). Especially in sheltered areas, the low weathering permits the accumulation of copper ions and enhancement in the acidity of water films [38].

A possible grouping according to the XRD results is: group A of the four B35 cakes where silicate minerals dominate, group B of the two B43 cakes which are mostly comprised of amorphous material, and finally group C of the red ferrous mineral powder which comprises entirely crystalline phases (Table 5). Gypsum ($CaSO_4 \cdot 2H_2O$) is a common constituent in groups B and C. It is known, in antiquity, for adhesive and quick binding properties, especially regarding constructions [27]. Especially and only in the B37-I bassanite [$CaSO_4 \cdot 0.5(H_2O)$], a semi-hydrous form of sulfate salt is detected. Bassanite probably resulted from the dehydration of gypsum under dried conditions. The authors, however, express their reservation that the presence of gypsum and bassanite in B37-I may be explained by contamination with reservation materials applied. In cake B35-I, the mineral hydrozincite $Zn_5(CO_3)_2(OH)_6$, which was used in antiquity for ophtalmic purposes, is detected [39]. The mineral cristobalite is only detected in the contents of B43a and B43b bowls. It is a member of the quartz group minerals and has the same chemical formula as quartz, SiO_2, but a distinct crystal structure. The above grouping of the contents under study implies a targeted selection of raw materials in order to have the desired effects during use.

In the sample B35-IIa, pentanoic acid has been detected (Table 6). Pentanoic acid or valeric acid is a compound with an unpleasant rancid odour and naturally occurs in the roots of the plant "valeriana officinalis". It is used in pharmaceuticals, as well as in cosmetics and perfumes, in its pure form or as an ester [40]. "Valeriana officinalis" was traditionally used in ancient Greece as a treatment against stress and insomnia [41]. At the same sample, another organic substance detected is nonadecane. Nonadecane is used as a fragrance agent and it is one of the components of the essential oils of some of the species that belong to the genus "Artemisia" [42]. More specifically, "Artemisia absinthium" is a species of Artemisia native to Eurasia and analysis has shown that, among several other chemical substances, its essential oil also contains nonadecane [43]. In ancient Greece, "Artemisia absinthium" was used for its medicinal properties against gastrointestinal disorders, as an astringent, helminthicide, diuretic, and emmenagogue [41]. In this case, nonadecane could have been used for its sole properties or as a masking agent to cover the unpleasant odour deriving from pentanoic acid. Nevertheless, access to other species of Artemisia or even completely different genuses, not native to the Mediterranean or Eurasia, cannot be excluded. As traces of fatty acid esters can be observed in all the compartments of the lidded case (B35), a possible explanation could be a preparation in the form of an ointment or even a formulation to be consumed orally, which would target the gastrointestinal system, simultaneously combining the sedative properties of "Valeriana officinalis" and the therapeutic activity of Artemisia absinthium, as gastrointestinal disorders are commonly associated with stress.

Hexadecanoic acid was detected in the B-35-I cake. This chemical substance is also known by the name palmitic acid and is used widely as a cleansing agent and in the production of soaps [44]. Palmitic acid can be found naturally in meat, dairy products, palm oil, and palm kernel oil, as well as in other plant oils, in smaller amounts [45]. More specifically, palmitic acid is one of the constituents of the essential oil of the plant "carum carvi", widely known as "caraway", which was used in Europe as a traditional medicine for stomach disorders and flatulence [41]. The content of sample B-35-I could potentially constitute a soap or a cleansing agent, some kind of medicinal ointment, or a decoction for oral consumption to treat stomach disorders.

It is well known that fatty acids and their derivatives were used in antiquity as constituents of pharmaceutical products [41,46]. According to the results of the HS-SPME/GC-MS analysis (Table 6), many of the twenty-two organic compounds that were detected in cake B35-IIa belong to the above classes. In the same cake, 6,6′-dibromoindigotin (DBI) was detected, which in combination with the reddish colour, denotes the existence of the royal dye shellfish purple. In the other cakes, very few organic compounds were detected. A possible interpretation is that cake B35-IIa was the basis that contained a large number of drastic substances and the reddish—purple colour a useful indicator for distinguishing it easily from other preparations. Alternatively, cake B35-IIa was perhaps contaminated by purple-dyed items that had been deposited in the burial. Dioscuridies Pedanius, in his work "De Materia Medica" [46], refers to the storage of fats in tin containers for pharmaceutical purposes. The metal case B35 and pyxis B37 were made of a high tin bronze alloy, but on the other hand, bowls B43a and B43b were made of almost pure Cu.

5. Conclusions

The combination of spectrometric (EDμXRF, XRD) and chromatographic (HPLC-DAD, HS SPME/GC-MS) analytical techniques was used as a holistic archaeometric approach in order to determine the construction technology of a group of metal containers and the chemical synthesis of their contents. The analytical methodology gave priority to a non-destructive implementation, as far as possible, due to the uniqueness and significance of the analyzed material. For the elemental analysis of the metal containers, EDμXRF spectrometry was implemented, in a non-invasive way, which resulted in the determination of their raw materials. Moreover, differentiations in their constituent parts were found. Generally, two kinds of alloy were used: a high tin bronze alloy for the construction of the main parts of metal case B35 and pyxis B37 and pure Cu accompanied by a low amount of impurities for the manufacture of bowls B43a and B43b. Furthermore, their chemical composition reveals the ancient archaeometallurgical procedures of the 4th century BC.

The implementation of a non-invasive technique like EDμXRF could be used as a guide for the selective implementation of a destructive one, e.g., the detection of Br in cake B35-IIa, which led to the implementation of HPLC-DAD and the detection of the royal dye shellfish purple. Although the archaeological approach led to the conclusion that this set was used for medical purposes [4], the archaeometric approach provides us with further information on medical knowledge in Macedonia during the 4th century BC

Author Contributions: Conceptualization, C.S.K., D.I.; Methodology, C.S.K., D.I., G.A.Z.; Investigation, C.S.K., A.Z., N.K., I.K.; Writing-Original Draft Preparation, C.S.K., D.I., N.K., I.K.; Writing-Review & Editing, C.S.K.; G.A.Z.; Supervision, G.A.Z.

Funding: This research received no external funding.

Acknowledgments: The authors wish to thank: P. Adam—Veleni, then Director of the Archaeological Museum of Thessaloniki (A.M.Th.) for the permission to analyze the artefacts of the article; V. Michalopoulou, Conservator of Antiquities, A.M.Th., for providing support with issues concerning ancient technology and metal conservation; and the two anonymous reviewers for their recommendations which resulted in a greatly improved paper.

Conflicts of Interest: The authors declare no conflict of interest.

References

1. Themelis, P.; Touratsoglou, G. *The tombs of Derveni*; Ministry of Culture–Archaeological Receipts Fund: Athens, Greece, 1977. (In Greek) (Θέμελησ, Π., Τουράτσογλου, Γ., 1997. Ο ι τάφοι του Δερβενίου, Υπουργείο Πολιτισμού—Ταμείο Αρχαιολογικών Πόρων και Απαλλοτριώσεων, Αθήνα)
2. Tsigarida, B.; Ignatiadou, D. *The Gold of Macedon: Archaeological Museum of Thessaloniki, Archaeological Receipts Fund*; Archaeological Museum of Thessaloniki: Thessaloniki, Greece, 2000.
3. Chrysostomou, P. Contributions to the history of medicine in ancient Macedonia. *Eulimene* **2002**, 3, 99–116. (In Greek) (*Χρυσοστόμου*, Π. 2002, «Συμβολέσ στην *Ιστορία της Ιατρικής στην Αρχαία Μακεδονία*», Ευλιμένη 3, 99–116»)
4. Ignatiadou, D. The warrior priest in Derveni grave B was a healer too, Histoire. *Méd. Santé* **2016**, *8*, 89–113. [CrossRef]
5. Guerra, M.F. Analysis of archaeological metals, The place of XRF & PIXE in the determination of technology and provenance. *X-ray Spectrom.* **1998**, *27*, 73–80.
6. Cechak, T.; Hlozek, M.; Musilek, L.; Trojek, T. X-ray fluorescence in investigations of archaeological finds. *Nucl. Instrum. Methods Phys. Res. Sect. B* **2007**, *B263*, 54–57. [CrossRef]
7. Karydas, A.G. Application of a portable XRF spectrometer for the non invasive analysis of museum metal artefacts. *Ann. Chim.* **2007**, *97*, 419–432. [CrossRef] [PubMed]
8. Charalambous, A.; Kassianidou, V.; Papasavvas, G. A compositional study of Cypriot bronzes dating to the Early Iron Age using portable X-ray fluorescence spectrometry (pXRF). *J. Archaeol. Sci.* **2014**, *46*, 205–216. [CrossRef]
9. Stuart, B. *Analytical Techniques in Materials Conservation*; John Wiley & Sons Ltd.: Chichester, UK, 2007; p. 194.
10. Bronk, H.; Rohrs, S.; Bjeoumikhov, A.; Langhoff, N.; Schmalz, J.; Wedell, R.; Gorny, H.-E.; Herold, A.; Waldschlager, U. Artax—A new mobile spectrometer for EDXRF spectrometry on art & archaeological objects, Fresenius's. *J. Anal. Chem.* **2001**, *371*, 307–316.
11. Dylan, S. Handheld XRF analysis of Renaissance bronzes: Practical approaches to quantification and acquisition. In *Handheld XRF for Art and Archaeology*; Shugar, A.N., Mass, J.L., Eds.; Leuven University Press: Leunen, The Netherlands, 2012; pp. 37–74.
12. Scott, D.A. *Ancient Metals: Microstructure and Metallurgy*; Lulu.com: Morrisville, NC, USA, 2010; Volume I.
13. Klug, H.P.; Alexander, L.E. *X-ray Diffraction Procedures for Polycrystalline and Amorphous Materials*; John Wiley & Sons: New York, NY, USA; Sydney, Australia; Toronto, ON, Canada, 1974; 966p.
14. Kantiranis, N.; Stergiou, A.; Filippidis, A.; Drakoulis, A. Calculation of the percentage of amorphous material using PXRD patterns. *Bull. Geol. Soc. Greece* **2004**, *36*, 446–453.
15. Vasileiadou, A.; Karapanagiotis, I.; Zotou, A. Determination of Tyrian purple by high performance liquid chromatography with diode array detection. *J. Chromatogr. A* **2016**, *1448*, 67–72. [CrossRef] [PubMed]
16. Karapanagiotis, I.; Mantzouris, D.; Cooksey, C.; Mubarak, M.S.; Tsiamyrtzis, P. An improved HPLC method coupled to PCA for the identification of Tyrian Purple in archaeological and historical samples. *Microchem. J.* **2013**, *110*, 70–80. [CrossRef]
17. Healy, J.F. *Mining and Metallurgy in the Greek and Roman World*; Thames and Hudson: London, UK, 1978.

18. Varoufakis, G. Technological specifications of the 4th century BC. Contribution to the historic metallurgy. *Archaeol. Ephemer.* **1974**, *113*, 57–62. (In Greek). (Βαρουφάκησ, Γ., 1974. Τεχνικαί προδιαγραφαί του 4^{ου} αι. π.Χ. Συμβολή εισ την ιστορικήν μεταλλουργίαν, Αρχαιολογική Εφημερίσ, σελ. 57–62)
19. Asimenos, K. Macedonian metalworking and alloy composition of the classical era. In Proceedings of the XII International Congress of Classical Archaeology, Athens, Greece, 4–10 September 1983; Volume C, pp. 283–288. (In Greek) (Ασημενόσ, Κ., 1988. Μακεδονική μεταλλοτεχνία και σύσταση κραμάτων τησ κλασικήσ εποχήσ, στο XII Διεθνέσ Συνέδριο Κλασσικήσ Αρχαιολογίασ, τόμοσ Γ', Αθήνα, σελ. 283–288)
20. Scott, D.A. *Metallography and Microstructure of Ancient and Historic Metals*; The Getty Conservation Institute: Singapore, 1991.
21. Ashkenazi, D.; Iddan, N.; Tal, O. Archaeometallurgical characterization of Hellenistic metal objects: The contribution of the objects for Rishon Le-Zion (Israel). *Archaeometry* **2012**, *54*, 528–548. [CrossRef]
22. Konečná, R.; Fintová, S. Copper and copper alloys: Casting, classification and characteristic microstructures. In *Copper Alloys—Early Applications and Current Performance—Enhancing Processes*; Collini, L., Ed.; InTech Web.Org: Rijeka, Croatia, 2012; pp. 3–30.
23. Tylecote, R.F. *The Early History of Metallurgy in Europe*; Longman Inc.: New York, NY, USA, 1987.
24. Craddock, P.T. The composition of the copper alloys used by the Greek, Etruscan and Roman civilisations: 2. the Archaic, Classical and Hellenistic Greeks. *J. Archaeol. Sci.* **1977**, *4*, 103–123. [CrossRef]
25. Craddock, P.T. The composition of the copper alloys used by the Greek, Etruscan and Roman Civilizations. *J. Archaeol. Sci.* **1976**, *3*, 93–113. [CrossRef]
26. Figueiredo, E.; Silva, R.J.C.; Senna-Martinez, J.C.; Araújo, M.F.; Braz Fernandes, F.M.; Inês Vaz, J. Smelting and recycling evidences from the Late Bronze Age habitat site of Baiões (Viseu, Portugal). *J. Archaeol. Sci.* **2010**, *37*, 1623–1634. [CrossRef]
27. Goffer, Z. *Archaeological Chemistry*, 2nd ed.; John Wiley & Sons, Inc.: Hoboken, NJ, USA, 2007.
28. Descamps- Lequime, S. The color of bronze—Polychromy and the aesthetics of bronze surfaces. In *Power and Pathos—Bronze Sculpture of the Hellenistic World*; Daehner, J.M., Lapatin, K., Eds.; J. Paul Getty Museum: Los Angeles, CA, USA, 2015; pp. 150–165.
29. Craddock, P.T.; Meeks, N.D. Iron in ancient copper. *Archaeometry* **1987**, *29*, 187–204. [CrossRef]
30. Henderson, J. *The Science and Archaeology of Materials—An Investigation of Inorganic Materials*; Routledge: New York, NY, USA, 2000.
31. Cooke, S.R.B.; Aschenbenner, S. The occurrence of metallic iron in ancient copper. *J. Field Arcahaeol.* **1975**, *2*, 251–266.
32. Papadimitriou, G. Copper and bronze metallurgy in ancient Greece. In *Archaeometry 90, Proceedings of the International Symposium in Archaeometry, 2–6 April 1990, Heidelberg, Germany*; Wagner, G.A., Pernicka, E., Eds.; Birkhauser Verlag: Basel, Switzerland, 1991; pp. 117–126.
33. Oudbashi, O.; Davami, P. Metallography and microstructure interpretation of some archaeological tin bronze vessels from Iran. *Mater. Charact.* **2014**, *97*, 74–82. [CrossRef]
34. Craddock, P.T.; Giumlia-Mair, A. Problems and possibilities for provenancing bronzes by chemical composition. In *Bronzeworking Centers of Western Asia, c.100-539 B.C., Kegan Paul International in Association with the British Museum*; Curtis, J., Ed.; Kegan Paul International in association with the British Museum; Distributed by Routledge; Chapman & Hall: London, UK, 1988.
35. Rovira, S.; Montero, I. Natural tin–bronze alloy in Iberian Peninsula metallurgy: Potentiality and reality. In *The Problem of Early Tin*; Giumlia-Mair, A., Schiavo, F.L., Eds.; Archaeopress: Oxford, UK, 2003; pp. 15–22.
36. Jenkins, R. *X-ray Fluorescence Spectrometry*; John Whiley & Sons Inc.: Hoboken, NJ, USA, 1999.
37. Gettens, R.J.; Stout, G.L. *Paintings Materials: A Short Encyclopedia*; Dover Publications: New York, NY, USA, 1966.
38. Leygraf, C.; Graedel, T.E. *Atmospheric Corrosion*; Wiley-Interscience: New York, NY, USA, 2000.
39. Giachi, G.; Pallecchi, P.; Romualdi, A.; Ribechini, E.; Lucejko, J.J.; Colombini, M.P.; Lippi, M.M. Ingredients of a 2000-y-old medicine revealed by chemical, mineralogical and botanical investigations. *PNAS* **2013**, *110*, 1193–1196. [CrossRef] [PubMed]
40. Goldberg, I.; Rokem, J.S. Organic and Fatty Acid Production, Microbial. In *Encyclopedia of Microbiology*, 3rd ed.; Elsevier: New York, NY, USA, 2009; pp. 421–442.

41. Skaltsa, E. *History of Pharmacy*; Kallipos-Hellenic Academic Books: Athens, Greece, 2015; p. 104. (In Greek) (Σκαλτσά, E., 2015. «Ιστορία τησ Φαρμακευτικήσ», Κάλλιποσ-Ελληνικά Ακαδημαϊκά Συγγράμματα και Βοηθήματα, Αθήνα, 104)
42. Mojarrab, M.; Delazar, A.; Esnaashari, S.; Afshar, F.H. Chemical composition and general toxicity of essential oils extracted from the aerial parts of Artemisia armeniaca Lam. and A. incana (L.) Druce growing in Iran. *Res. Pharm. Sci.* **2013**, *1*, 65–69.
43. Kordali, S.; Cakir, A.; Mavi, A.; Kilic, H.; Yildirim, A. Screening of Chemical Composition and Antifungal and Antioxidant Activities of the Essential Oils from Three Turkish Artemisia Species. *Agric. Food Chem.* **2005**, *53*, 1408–1416. [CrossRef] [PubMed]
44. Anneken, D.J.; Both, S.; Christoph, R.; Fieg, G.; Steinberner, U.; Westfechtel, A. Fatty Acids. In *Ullmann's Encyclopedia of Industrial Chemistry*; Wiley-VCH: Weinheim, Germany, 2006.
45. Kim, S.; Thiessen, P.A.; Bolton, E.E.; Chen, J.; Fu, G.; Gindulyte, A.; Han, L.; He, J.; He, S.; Shoemaker, B.A.; et al. PubChem Substance and Compound databases. *Nucleic Acids Res.* **2016**, *44*, D1202–D1213. [CrossRef] [PubMed]
46. Dioscorides, P. *De Materia Medica*; Beck, L.Y., Translator; Altertumswissenschaftliche Texte und Studien, Band 38; Olms-Weidmann: Hildesheim, Germany, 2005.

Review

Food Sample Preparation for the Determination of Sulfonamides by High-Performance Liquid Chromatography: State-of-the-Art

Dimitrios Bitas [1], Abuzar Kabir [2], Marcello Locatelli [3] and Victoria Samanidou [1,*]

1 Laboratory of Analytical Chemistry, Department of Chemistry, Aristotle University of Thessaloniki, 54124 Thessaloniki, Greece; dimitriosbitas@gmail.com

2 International Forensic Research Institute, Department of Chemistry and Biochemistry, Florida International University, 11200 SW 8th St, Miami, FL 33199, USA; akabir@fiu.edu

3 Department of Pharmacy, University "G. d'Annunzio" of Chieti-Pescara, 66100 Chieti, Italy; m.locatelli@unich.it

* Correspondence: samanidu@chem.auth.gr; Tel.: +30-231-099-7698

Received: 30 April 2018; Accepted: 28 May 2018; Published: 4 June 2018

Abstract: Antibiotics are a common practice in veterinary medicine, mainly for therapeutic purposes. Sectors of application include livestock farming, aquacultures, and bee-keeping, where bacterial infections are frequent and can be economically damaging. However, antibiotics are usually administered in sub-therapeutic doses as prophylactic and growth promoting agents. Due to their excessive use, antibiotic residues can be present in foods of animal origin, which include meat, fish, milk, eggs, and honey, posing health risks to consumers. For this reason, authorities have set maximum residue limits (MRLs) of certain antibiotics in food matrices, while analytical methods for their determination have been developed. This work focuses on antibiotic extraction and determination, part of which was presented at the "1st Conference in Chemistry for Graduate, Postgraduate Students and PhD Candidates at the Aristotle University of Thessaloniki". Taking a step further, this paper is a review of the most recent sample preparation protocols applied for the extraction of sulfonamide antibiotics from food samples and their determination with high-performance liquid chromatography (HPLC), covering a five-year period.

Keywords: food; sample preparation techniques; sulfonamides; high-performance liquid chromatography; HPLC; ultra-high-performance liquid chromatography; UHPLC

1. Introduction

The scope of this state-of-the-art review is to cover the literature regarding the sample preparation protocols developed for the extraction of sulfonamides (SAs) from food samples followed by high-performance liquid chromatography (HPLC) determination, covering the last five years. The review is divided in two main sections. In this first main section, a theoretical background on veterinary drugs and antibiotic use, sulfonamides and their applications, the reported sample preparation techniques, as well as the chemical composition of the reported food matrices and official methods for the determination of antibiotics/sulfonamides in foodstuff samples are provided for the reader to have a prompt introduction to basic terminology. More details can be found in the cited review articles and book sections. In the second main section, the reported sample preparation protocols are provided in full detail for each food matrix.

1.1. Veterinary Drugs—Antibiotics

The European Union Council Directive 96/23/EC "on measures to monitor certain substances and residues thereof in live animals and animal products" divides all pharmacological substances used for veterinary purposes and their corresponding residues into two main groups, A and B. Group A includes substances with anabolic effect, such as antithyroid agents, steroids, and β-agonists, as well as unauthorized substances, such as chloramphenicol, chlorpromazine, metronidazole, and nitrofurans. Group B includes veterinary drugs, such as antibiotics, anthelmintics, anticoccidials and non-steroidal anti-inflammatory drugs, and environmental contaminants, such as organochlorine and organophosphorus compounds [1].

Antibiotics are used in veterinary medicine in order to improve the health and increase the productivity of the food-producing animals and at the same time reduce the morbidity and mortality rates among the livestock. Animals are administered with antibiotics not only for therapeutic purposes but also as prophylactic and metaphylactic measures [2]. Prophylaxis is a preventative measure, where animals are administered with sub-therapeutic doses and in some cases full doses of antibiotics through feed or water and is a common practice in massive livestock production. Metaphylaxis is a measure taken when a number of animals exhibits some of the disease symptoms, and all the animals are administered in order to prevent the disease from spreading. However, both practices are not always effective due to the fact that some antibiotic groups are active during bacterial cell proliferation [3]. The most common routes of administration are through water and feed medication, followed by injection [2]. Antibiotics can also be used as feed additives. Their intake favors the natural intestinal flora of the animals by inhibiting the harmful microorganisms, and as a result, nutrient absorption and assimilation are increased, thus providing a growth promoting effect [3]. Aminoglycosides, β-lactams (penicillins and cephalosporins), macrolides, phenicols, quinolones, SAs, and tetracyclines are among the most common antibiotic substances in veterinary medicine [4].

The presence of veterinary drug residues in animal-originated products relies on the physicochemical properties of each compound that affect the absorption, distribution, metabolism, and excretion of the drug from the animal body. For this reason, a withdrawal period, between the last drug administration and the slaughter or milk, egg, and honey collection, is established in order to ensure the existence of animal-originated products with low drug residues [4]. However, the extensive use of antibiotics besides therapeutic purposes has led to the presence of antibiotic residues in animal originated food products and the development of bacterial drug resistance [3]. The main reasons for antibiotic residues is the illegal use and uncontrolled administration of veterinary drugs to healthy animals, as well as not taking into consideration the withdrawal period, the lack of professional advice, extra label use, frequent drug administration, and increased dosing, as well as contaminated housing, water, and animal feed [4–6]. Antibiotic residues can cause adverse effects to consumers, such as acute allergic or toxic reactions, chronic toxic effects from prolonged exposure to antibiotic residues, and natural intestinal flora disruption [5].

Liver, kidney, and fat are animal tissues with high antibiotic residue concentration, while muscle tissue has relatively lower residues [5]. The European Union Commission Regulation No 37/2010 sets the legislative basis for the "pharmacologically active substances and their classification regarding maximum residue limits in foodstuffs of animal origin", by classifying the substances into the categories of allowed and prohibited. For the allowed substances, a maximum residue limit (MRL) is provided for each target tissue expressed in micrograms per kilogram (μg/kg) of fresh target tissue. Target tissues include muscle, fat, liver, and kidney, as well as milk, eggs, and, in a few cases, honey. The MRLs usually refer to a specific compound, a metabolite, or a compound mixture [7].

1.2. Sulfonamides

SAs constitute a wide-spectrum synthetic antibiotic category effective against a wide range of bacterial species, such as *Bacillus* spp., *Brucella* spp., *Streptococcus* spp., staphylococci gram-positive aerobes, and enterobacteriaceae gram-negative aerobes, as well as protozoa, parasites, and fungi. SAs are derivatives of sulfanilamide and the various SA analogues, that result from the various **R** radicals of the $-SO_2NH\mathbf{R}$ group (Figure 1). Each SA analogue has different physicochemical and pharmacokinetic properties and as a result antibiotic effect. SAs are relatively insoluble with solubility increasing in alkaline pH, which is an important factor for the type of administration and disease treatment they are intended for [8].

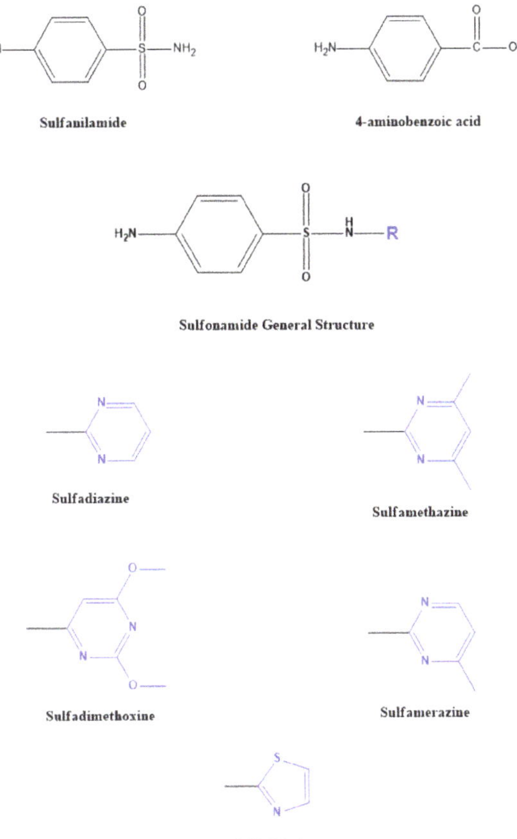

Figure 1. Diagrams of the chemical structure of sulfonamide, sulfanilamide, 4-aminobenzoic acid, and the chemical structure of the most common SA analogues in veterinary medicine.

The SA antibacterial effect relies on their ability to inhibit the conversion of folic acid to tetrahydrofolic acid by competitively antagonizing 4-aminobenzoic acid (Figure 1) for the dihydropteroate synthetase enzyme. Tetrahydrofolic acid is essential for the nucleic acid synthesis, thus SAs inhibit the bacterial DNA and RNA synthesis and subsequently the protein synthesis. Additionally, SAs reduce the bacterial cell permeability for glutamic acid that is essential for the folic acid synthesis. SAs are effective against bacterial species that synthesize the required 4-aminobenzoic acid, but ineffective in the presence of increased amounts of 4-aminobenzoic acid and against species that receive the required 4-aminobenzoic acid from other sources or have antibiotic resistance [6].

SA formulations usually consist of a SA analogue and a diaminopyrimidine, such as aditoprim, baquiloprim, ormetoprim, or trimethoprim, that have a synergistic interaction. The liver and kidney are animal tissues with the highest concentrations of SAs and their metabolites [4]. Sulfamethoxazole, sulfacetamide, and sulfasalazine are SA analogues commonly used in human medicine, while sulfadiazine (SDZ), sulfamethazine (SMZ), sulfadimethoxine (SDMX), sulfamerazine, and sulfathiazole are commonly used in veterinary medicine [6]. The European Union Commission

Regulation No 37/2010 sets the MRL for all SA analogues to 100 μg/kg for muscle, fat, liver, and kidney from all food-producing species and bovine, ovine, and caprine milk, while their use is prohibited for animals that produce eggs for human consumption. In the case of more than one SA analogue, the sum of the SA residues should not exceed the provided MRL value (Table 1) [7]. Honeybees are also considered food-producing species and antibiotics are used in beekeeping in order to treat two of the most severe diseases for bees, the American/European foulbrood, and nosemosis. More specifically, the SA analogue sulfothiazole is used against the American foulbrood caused by *Paenibacillus larvae* in order to prevent the spread of the disease, suppress the symptoms, and inhibit the spore germination. SAs can also be used prophylactically against nosemosis. However, a beehive lacks in metabolic pathways for the elimination of antibiotic residues and a withdrawal period does not apply in the case of bees. For this reason, the European Union has not established MRLs for antibiotic residues in honey and only veterinary drugs with zero residues are authorized for beekeeping [9].

Table 1. MRLs for sulfonamide antibiotics, provided by the EU, Codex Alimentarius and FDA.

Sulfonamide	Animal Species	MRL	Target Tissue
Commission Regulation (EU) No 37/2010 [7]			
Sulfonamides	All food producing species	100 μg/kg	Muscle Fat Liver Kidney
	Bovine Ovine Caprine	100 μg/kg	Milk
Codex Alimentarius [10]			
Sulfadimidine	Cattle	25 μg/L	Milk
	Not specified	100 μg/kg	Muscle Liver Kidney Fat
CFR—Code of Federal Regulations—U.S. Food & Drug Administration (FDA) [11]			
Sulfabromomethazine sodium	Cattle Not specified	100 μg/kg 10 μg/L	Uncooked edible tissue Milk
Sodium sulfachloropyrazine monohydrate	Chicken	0	Uncooked edible tissue
Sulfachlorpyridazine	Calves Swine	100 μg/kg	Uncooked edible tissue
Sulfadimethoxine	Chickens Turkeys Cattle Ducks Salmonids Catfish Chukar partridges	100 μg/kg	Uncooked edible tissue
	Not specified	10 μg/L	Milk
Sulfaethoxypyridazine	Cattle Swine Not specified	100 μg/kg 0 0	Uncooked edible tissue Uncooked edible tissue Milk
Sulfamerazine	Trout	0	Uncooked edible tissue
Sulfamethazine	Chickens Turkeys Cattle Swine	100 μg/kg	Uncooked edible tissue
Sulfaquinoxaline	Chickens Turkeys Calves Cattle	100 μg/kg	Uncooked edible tissue

The most crucial problem arising from the uncontrolled use of SAs is the development and spreading of drug resistance, rather than the presence of SA residues in animal originated products itself. SA drug resistance derives from mutations in the dihydropteroate synthase gene that results in enzymes with structural alterations with decreased affinity towards the SAs. Drug resistance genes can be transferred between bacterial strains or genera during a horizontal gene transfer in plasmids, transposons, or integrons. For this reason, SA resistant strains are higher than tetracyclines and other antibiotics, while bacterial species with more than one gene for SA resistance have been observed. Furthermore, drug resistance in an antibiotic group can favor the development of cross-resistance [6].

1.3. Sample Preparation

Food samples are complex heterogenous matrices, where all analytes are distributed in a random manner. Food analysis involves sampling, homogenization, and sample preparation that increase the analytical accuracy and precision. Focusing on sample preparation it usually involves storage, particle size reduction, homogenization, weighting, dilution, filtration, extraction, clean-up, and derivatization. Proper sample preparation protocols result in matrix interference elimination and analyte preconcentration, thus affecting the selectivity, sensitivity, detection capability, and the overall performance of an analytical technique. The most time-consuming step in analytical method development is the optimization of the sample preparation protocol that includes analyte extraction and clean-up. Some of the most common sample preparation techniques used in food analysis are liquid-liquid extraction (LLE), solid-liquid extraction (SLE), solid-phase extraction (SPE), solid-phase microextraction (SPME), stir bar sorptive extraction (SBSE), supercritical fluid extraction (SFE), accelerated solvent extraction (ASE), ultrasonic, and Soxhlet extraction [12].

1.3.1. LLE

LLE is one of the most used sample preparation techniques, along with SPE, and probably the oldest. In LLE the analytes are extracted from an aqueous sample into a water immiscible solvent, according to relative solubility. Despite the wide use of LLE, there are many disadvantages, such as increased time and solvent requirements, analyte lose, sample contamination, and low sensitivity, that reduce LLE applications in modern analytical chemistry. Liquid-phase microextraction (LPME) is an LLE-based microextraction technique that has been developed in order to overcome LLE disadvantages. Extraction takes place between the aqueous sample phase (donor phase) and a water immiscible solvent extraction phase (acceptor phase). LPME can be divided into single-drop microextraction (SMDE), hollow fiber liquid-phase microextraction (HF-LPME), and dispersive liquid-liquid microextraction (DLLME). DLLME is a ternary extraction system that involves the aqueous sample, an extraction solvent, and a disperser solvent. Microliters of the organic extraction solvent and the dispersive solvent are injected into the sample solution and extraction solvent droplets are formed with the help of the dispersive solvent, resulting in a cloudy solution. After the extraction equilibrium between the sample and the extraction solvent is achieved, the cloudy solution is centrifuged and the lower organic phase is collected for analysis. The extraction solvent should be water immiscible, such as chloroform, carbon tetrachloride, and dichloromethane, while the dispersive solvent should be miscible in both aqueous solution and organic solvent, such as ethanol, methanol, acetonitrile, and acetone. Following the principles of green analytical chemistry, conventional extraction solvent can be replaced by ionic liquids that are liquid organic salts, combinations of organic cations with inorganic anions with unique physicochemical properties. Ionic liquids are characterized by low volatility and high density, thus forming more stable droplets and better phase separation than the conventional organic solvents [13].

1.3.2. SLE

The principals of SLE are similar to LLE, except that the analytes are extracted from solid samples. The solid sample is mixed with extraction solvent, the two phases interact, and the soluble sample components diffuse into the extraction phase. Various organic solvents can be used in SLE, however, SLE is usually laborious and has analogous disadvantages as LLE, such as increased solvent requirements, partial extraction, solvent impurities, and emulsion formation. In order to increase the extraction efficiency of the organic solvent, heating and pressure or ultrasounds can be applied during the extraction. In pressurized liquid extraction (PLE), also known as ASE, the solid sample and the extraction solvent are transferred into an extraction cell and the extraction takes place under high temperature (40–200 °C) and pressure (500–3000 psi) for 5–15 min. After the extraction is complete, the sample extract is collected and purged. The increased temperature and pressure applied in PLE result in reduced extraction time, enhanced analyte's solubility and mass transfer, better solvent penetration, and overall improved extraction yields. However, PLE requires expensive equipment and high temperature solvent reduces the extraction selectivity and applicability of PLE for the extraction of less thermal stable compounds. Ultrasound-assisted extraction (UAE) is a less extreme extraction approach in which extraction is assisted by the application of ultrasounds. Compared with other sample preparation techniques, UAE is relatively faster with reduced solvent requirements, and at the same time enables the extraction of analytes in room temperature. However, this technique lacks selectivity and enrichment capability when extraction of trace amounts is required and is usually combined with other clean-up techniques for improved extraction efficiency [14].

1.3.3. Salting-Out Extraction

The salting-out effect has been exploited in sample preparation techniques, such as Quick Easy Cheap Effective Rugged Safe (QuEChERS) and salting-out liquid-liquid extraction (SALLE), in order to improve analyte's extraction from aqueous samples. The salting-out effect is based on the decrease of solubility of water soluble organic analytes when salt concentration in the aqueous sample solution is increased. This effect favors the partition of the analytes into a water-miscible organic solvent and separation between the two phases. Various water-miscible solvents can be used, but acetonitrile is the most convenient water-miscible organic solvent for the application of the salting-out effect due to being chemically inert with organic analytes and the most common mobile phase component in liquid chromatography, and at the same time, acetonitrile has the ability to precipitate matrix proteins. Salting-out agents are usually inorganic and organic salts that provide cations (Mg^{2+}, Sr^{2+}, Ca^{2+}, Ba^{2+}, K^+, Na^+, NH_4^+, Li^+) and anions (SO_4^{2-}, CH_3COO^-, Cl^-, NO_3^-, Br^-, I^-, CNS^-), such as $MgSO_4$, NaCl, CH_3COONH_4, CH_3COONa, and $(NH_4)_2SO_4$, which promote the transfer of the hydrophilic compounds to the organic phase. These salts should be soluble in the aqueous sample but have negligible solubility in the organic phase [15]. In either QuEChERS and SALLE procedures, a water-miscible solvent and a salting-out agent are sequentially added to the sample and the mixture is shaken and centrifuged. The formed organic phase is then collected and can be either injected directly for analysis or further treated for water removal and analyte isolation [14]. Aqueous two-phase system (ATPS) extraction is another alternative based on the partitioning of the analytes between two phases and is used for the extraction of analytes from aqueous samples. The most common biphasic system in food analysis is polymer/salt, while ionic liquid/salt systems were also reported. Polyethylene glycol is usually the polymer of choice combined with phosphate, sulfate, or citrate salts. Parameters that affect the extraction include the molecular weight and concentration of the polymeric phase, the solution pH, and temperature. ATPS does not require the use of organic solvents in contrast with other sample preparation techniques, while sample deproteinization or defatting is not always required [16].

1.3.4. SPE

SPE is a widely used sample preparation technique that provides high enrichment factors and recoveries with reduced sample volume requirements and automation capability. SPE can be used for the daily laboratory routine and in many cases SPE has replaced LLE. In SPE the components of the sample solution interfere with a solid phase (sorbent material) and separation between the target analytes and the matrix interferences can be achieved. The sorbent is usually packed into SPE cartridges between two frits. SPE cartridges are commercially available for specific applications or can be manually prepared with the selected sorbents. In a typical SPE procedure, the SPE cartridges are conditioned/activated with a solvent or a solvent mixture, the sample solution is loaded into the cartridge, the loaded cartridge is washed to remove the retained interreferences, and the retained analytes are eluted with an appropriate solvent (Figure 2a). However, SPE applications are restricted by the type of the sorbent and the characteristics of the sample components, while the tightly packed SPE cartridges increase the extraction time and cause backpressure [13]. These problems can be eliminated with dispersive solid-phase extraction (DSPE) where the sorbent is dispersed in the sample solution and not packed into a cartridge. After the dispersion, the solution is shaken, and when the extraction is complete, the sorbent is collected by centrifugation or filtration and the analytes are eluted by repeating the previous step with the use of an appropriate solvent (Figure 2b) [17]. C18 and OASIS® HLB are two of the most widely applied commercially available materials for SPE. C18 is a nonpolar sorbent that consists of octadecylsilane bonded to silica particles and is suitable for the reversed-phase binding of hydrophobic analytes. OASIS® HLB is a water-wettable hydrophilic-lipophilic-balanced polymeric reversed-phase sorbent that consists of N-vinylpyrrolidone and divinylbenzene and can be used for the binding of acidic, basic, or neutral analytes [14]. Novel SPE sorbents include carbon nanotubes (CNTs) [13], molecularly imprinted polymers (MIPs) [18], and magnetic materials. Magnetic solid-phase extraction (MSPE) is a SPE-based technique that employs magnetic sorbent materials. The magnetic materials comprise of a magnetic metal oxide nanoparticle core, usually Fe_3O_4, coated with inorganic or organic materials, such as silica, alumina, chitosan, or polypyrrole, while the coating can be modified with functional groups for improved sorption capability. A MSPE application is similar to DSPE, with the difference that the sorbent can be collected by means of a magnet (Figure 2b) [19]. While solid-phase techniques can be used for aqueous samples or sample solutions, matrix solid-phase dispersion (MSPD) can be applied directly for the extraction of analytes from solid, semi-solid, and viscous samples. Typically, the sample is mechanically blended with solid support in order to achieve matrix disruption and the development of interactions between the analytes and the sorbent material. The mixture is then transferred and packed into a SPE cartridge and the analytes are eluted with an appropriate sorbent. Apart from the applicability on solid samples, MSDP provides a simple and selective approach for sample extraction and clean-up in a single step [14].

Other solid-phase extraction variations include SPME and SBSE. In both techniques, the sorbent material is coated on a substrate, and extraction/analyte desorption follow the same principles. SPME employs silica or stainless-steel fibers coated with the sorbent material that can be used for the extraction of analytes from gaseous, liquid, and solid samples. A typical SPME procedure involves the partitioning of the analytes between the sorbent and the sample matrix and analyte desorption directly into the analytical instrument. SPME can be coupled with HPLC by means of a six-port injector combined with a desorption chamber, and desorption can be performed with an organic solvent or the mobile phase in static or dynamic mode. SMPE fibers can be coated with various sorbent materials, thus SMPE applicability can be expanded for the extraction of a wide range of analytes and sample matrices [14]. SBSE employs magnetic stir bars coated with polydimethylsiloxane that is a polar polymeric material and develops hydrophilic interactions with the analytes, such as hydrogen bonds and van der Waals forces. In a typical SBSE procedure, a coated stir bar is introduced into the sample solution and the analytes are absorbed onto the coating by continuous stirring. After the extraction, the bar is collected, washed with deionized water, and dried, while analytes can be desorpted, either thermally by thermal or liquid desorption [13]. Both SPME and SBSE are less time-consuming and have reduced sample and solvent requirements in comparison with SPE.

1.3.5. FPSE

Fabric phase sorptive extraction (FPSE) is a recently introduced novel microextraction technique that employs reusable cellulose or polyester fabric substrates homogenously coated with sol-gel hybrid sorbents. In a typical FPSE procedure, the coated fabric, along with a magnetic stir bar, are introduced into a vial that contains the sample or sample solution, and analyte extraction is conducted under stirring. The coated fabric is collected and placed for 4–10 min into a second vial that contains the eluting solvent (Figure 2c). The eluate can be centrifuged prior to analysis. The coated fabric is washed with an appropriate organic solvent and rinsed with deionized water between extractions. Reported FPSE coatings include polydimethylsiloxane, poly(ethyleneglycol), C18, and graphene. The FPSE protocol is simple and fast, with reduced solvent requirements, while the coated fabric can be introduced directly to the liquid samples and is compatible with a wide range of organic solvents. FPSE substrates are characterized by increased sorbent loading and improved adsorption capacity in comparison with SPME, providing high analyte preconcentration [20].

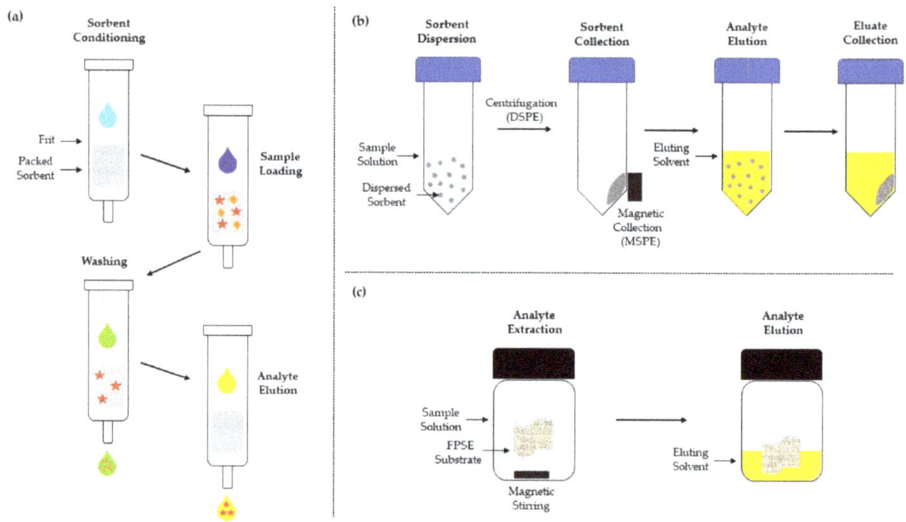

Figure 2. Schematic representation of the basic steps in (**a**) SPE, (**b**) DSPE and MSPE, and (**c**) FPSE.

1.4. Food Composition

Meat contains 72–75% water, 19% proteins, 2.5–5% lipids, 1% vitamins and carbohydrates, and 1% ash. Lipids can vary between 1% and 15%. The main edible animal tissue is the muscle, while other edible parts include organs, such as liver and kidneys, fat, and blood. Edible animal livestock species include bovine and porcine species and poultry. Animal lipids are mostly deposited under the skin (subcutaneous fat), between the muscles (intermuscular fat), and around organs, such as kidneys and heart, but varies between animal species [21]. The chemical composition of fish varies among species, with 50–60% of fish weight being muscle. Fish muscle contains 52–82% water, 16–21% proteins (or 10–25% for farmed species), 0.5–2.3% lipids, 1.2–1.5% ash, and 0.5% carbohydrate content. In comparison with red meat, fish lipid content is lower and ranges between 0.2% and 30%, while lipids are deposited in the liver, muscle, perivisceral, and subcutaneous tissues. Fatty fish species deposit fat all over the muscle tissue that is colored grey, yellow, or pink, such as salmon, and more than 50% of the skin consists of lipids (in other species it ranged between 0.2% and 3.9%) [22]. Milk is a heterogenous mixture that consists of water, emulsified fat, caseins and whey proteins, lactose, minerals, and vitamins. The gross cow milk composition is 86.3% water, 4.9% fat, 3.4% proteins, 4.1% lactose, and 0.7% ash. Buffalo, sheep, and goat milk can also be consumed by humans. Fat is the most important milk component, both organoleptically and commercially, and ranges from below 3% to more than 6% [23]. Eggs consist of white and yolk for 60% and around 30–33% of the total egg weight, respectively. The egg white or albumen is a protein solution that contains over 40 different proteins, with ovalbumin constituting the 54% of the total white proteins. The egg yolk contains lipoproteins, with low-density lipoproteins constituting the 65% of the total yolk proteins. Egg fat is mainly present in egg yolk as triacylglycerol and phospholipids and comprises the 9–10% of the total egg weight. Other egg components include minerals and vitamins that are also located in the egg yolk [24].

1.5. Official Methods of Analysis

Official methods for the determination of antibiotics/sulfonamides in food samples are available by regulating agencies. The website of the FDA provides a "Laboratory Methods—Drug & Chemical Residues Methods" section [25] that includes analytical methods for the determination of multiple phenicol residues in honey samples with electrospray liquid chromatography—mass spectrometry (LC-MS) and chloramphenicol residues in crustacean species (shrimp, crab, crawfish) samples with liquid chromatography—tandem mass spectrometry (LC-MS/MS) [26], as well as the determination of fluoroquinolone residues in milk samples with LC-MS/MS [27]. Two multi-class and multi-residue LC-MS/MS methods for the determination of drug residues in milk and aquaculture samples (fish, shrimp) are provided in the "Field Science and Laboratories—Laboratory Information Bulletins" section of the FDA website [28]. A high-performance liquid chromatography—ultraviolet detection (HPLC-UV) method for the determination of sulfamethazine in milk samples is also provided by the FDA [29]. The Association of Official Analytical Chemists (AOAC) provides the "AOAC Official Method 993.32" for the determination of eight sulfonamide residues in raw bovine milk samples with liquid chromatography—ultraviolet detection (LC-UV) [30].

2. Extraction of Sulfonamides from Food Samples

Meat, milk, eggs, and honey are animal-originated products with increased demand. In this section the detailed sample preparation protocols are provided for each reported paper.

2.1. Animal Tissue Samples

2.1.1. SLE

The same team developed two SLE protocols for multi-class antibiotic extraction, including 15 SAs, from bovine tissue [31] and fish tissue samples [32] followed by ultra-high-performance liquid chromatography—mass spectrometry (UHPLC-MS/MS). In the first protocol, spiked bovine tissue (2 g) was mixed with acetonitrile (ACN) (10 mL) and ethylenediaminetetraacetic acid (EDTA) (0.1 M, 1 mL) and the mixture was shaken for 20 min and centrifuged at 3100 *g* for 15 min. The supernatant was collected, mixed with *n*-hexane (3 mL), vortexed for 30 s, centrifuged at 3100× *g* for 15 min, and the *n*-hexane layer was discarded. The extract was evaporated under nitrogen stream at 40 °C to 0.5 mL final volume, dissolved in mobile phase (400 μL), and passed through a 0.45 μm filter, prior to analysis. The authors emphasized the simplicity, as well as the reduced cost and time requirements of the developed sample preparation protocol. The addition of EDTA, the sample defatting with *n*-hexane, and the evaporation to 0.5 mL were employed in order to increase the recoveries of all analytes [31]. The same protocol was applied for the fish tissue samples, except for the *n*-hexane treatment step. Spiked fish tissue (2 g) was mixed with ACN (10 mL) and EDTA (0.1 M, 1 mL), and the mixture was shaken for 20 min and centrifuged at 3100× *g* for 15 min. The supernatant was collected and evaporated under nitrogen stream at 40 °C and the residue was dissolved in mobile phase (400 μL) [32]. In both cases, a SPE sample clean-up step was omitted due to increased selectivity over specific antibiotics that prohibits multi-class extraction, while the sample preparation protocol cost and time requirement were decreased and the number of samples analyzed in a daily routine was increased. Another SLE protocol was reported for the extraction of SDZ, SMZ, SIX, SDMX, and sulfaquinoxaline (SQX) from shrimp tissue samples. Spiked tissue (0.5 g) was mixed with methanol (MeOH)-ACN (50:50, *v*/*v*; 1 mL) and the mixture was vortexed, sonicated for 15 min, and centrifuged at 3500 rpm for 10 min. The extraction step was repeated twice with MeOH-ACN (50:50, *v*/*v*; 1 mL) and twice with 0.1% CH_3COOH-MeOH (60:40, *v*/*v*; 0.5 mL) and the supernatants were collected between extractions. All collected supernatants were combined, evaporated to dryness under nitrogen stream, and the residue was dissolved in MeOH (500 μL) and passed through a 0.20 μm syringe filter. Analysis was carried out by high-performance liquid chromatography—diode array detection (HPLC-DAD). SPE and MSPD were also tested but higher recoveries were achieved with the developed sample preparation protocol [33].

Two ASE protocols were reported for the extraction of 15 SAs and metabolites from baby food samples [34], and multi-class antibiotic extraction, including 9 SAs, from fish tissue samples [35]. In the first protocol, spiked baby food sample (5 g) was transferred into an ASE extraction cell and mixed with 1% CH_3COOH in MeOH-ACN (4:1, *v*/*v*). Extraction was carried out at 70 °C and 1500 psi with 5 min preheating time and 5 min static time, while flush volume was 60% and purge time 60 s. Extraction was repeated for 3 cycles. The extracts (3 × 18 mL) were diluted with extraction solvent (3 × 20 mL), placed in the freezer at −18 °C overnight and centrifuged at 4 °C and 4000 rpm for 5 min. The supernatant was collected and passed through a 0.45 μm filter and analyzed with ultra-high-performance liquid chromatography—orbitrap high-resolution mass spectrometry (UHPLC-Orbitrap-MS). The developed sample preparation protocol provided higher recoveries in comparison with an official AOAC QuEChERS extraction protocol [34]. In the second report, the fish tissue samples were purified with C18 resin inside the ASE extraction cell. The sample extracts were evaporated to dryness, dissolved in mobile phase, and centrifuged at −4 °C and 10,000 rpm prior to high-performance liquid chromatography—mass spectrometry (HPLC-MS/MS) analysis. However, information could be collected only from the abstract and the further details are not available because the rest of the paper is written in Chinese [35].

Two extraction protocols, PLE and USE, were reported for the extraction of 16 SAs and metabolites from chicken, sheep, fish, and horse tissue samples. In both cases, analysis was carried out by high-performance liquid chromatography—quadrupole linear ion trap—mass spectrometry (HPLC-QqLIT-MS/MS). For the PLE protocol, spiked tissue (5 g) was defatted with hexane, mixed with diatomaceous earth, and transferred inside a PLE cell. The extraction solvent was 0.2% CH_3COOH in ACN, the preheating period was 8 min, and extraction was achieved at 90 °C and 1500 psi for 7 min. The total flush volume was 80%, 60 s of purging under nitrogen stream was applied and the extraction step was repeated three times. The collected extracts were placed in the freezer at −18 °C for 1 h and centrifuged at 3500 rpm for 10 min in order to remove the tissue proteins. The supernatant was evaporated to dryness under nitrogen stream at 40 °C and the residue was dissolved in H_2O-ACN (85:15, *v*/*v*; 1 mL). For the UAE protocol, spiked tissue (5 g) was mixed with ACN (10 mL) and the mixture was vortexed for 10 s and placed inside an ultrasonic bath for 60 min. Then, the mixture was placed in the freezer at −18 °C for 1 h and centrifuged at 3500 rpm for 10 min, in order to remove the tissue proteins. The supernatant was collected and evaporated to dryness under nitrogen stream at 40 °C. The residue was dissolved in H_2O-ACN (85:15, *v*/*v*; 2 mL), hexane (2 mL) was added, and the mixture was vortexed for 5 s and centrifuged at 3500 rpm for 10 min. The lower phase was used for analysis. Although both developed protocols could efficiently extract the SAs from the tissue samples, UAE is simpler and less solvent and time-consuming [36].

2.1.2. Salting-Out Extraction

Three QuEChERS extraction protocols were reported in the literature. In the first report, a QuEChERS extraction protocol was developed for the extraction of 22 SAs and metabolites from bovine, chicken, pork, and sheep tissue samples. Spiked tissue (5 g) was mixed with H_2O (5 mL), vortexed for 1 min, and 1% CH_3COOH in ACN (10 mL) was added to the mixture. Sodium hydrogen citrate sesquihydrate (0.5 g), sodium citrate (1 g), $MgSO_4$ (4 g), and NaCl (1 g) were added in sequence and the mixture was shaken and vortexed for 1 min between each addition. The mixture was centrifuged at 3500 rpm for 5 min, the supernatant (6 mL) was collected, mixed with primary secondary amine (PSA) (150 mg) and anhydrous $MgSO_4$ (900 mg), and the mixture was shaken, vortexed, and centrifuged as described. The supernatant (4 mL) was collected and evaporated to dryness under nitrogen stream at 35 °C. The residue was dissolved in 0.01% formic acid in 5% MeOH aqueous solution (500 μL), passed through a 0.22 μm nylon filter, and centrifuged at 13,500 rpm for 10 min prior to high-performance liquid chromatography—high-resolution mass spectrometry (HPLC-HRMS) analysis [37]. In the second report, a modified QuEChERS extraction protocol was developed for the extraction of eight SAs from chicken muscle and egg sample. The extraction protocol employed a commercially available sorbent material Z-Sep$^+$, consisting of C18 and zirconia both bonded to silica particles, for the chicken samples and PSA for the egg samples. Spiked tissue (5 g) was mixed with H_2O (5 mL) and 1% CH_3COOH in ACN (10 mL) and the mixture was shaken for 10 min. $MgSO_4$ (4 g) and CH_3COONa (1 g) were added and the mixture was shaken for 1 min, vortexed for 2 min, and centrifuged at 5000 rpm for 5 min. The supernatant (3 mL) was collected and mixed with C18-zirconia sorbent (300 mg) and the mixture was shaken for 30 s and vortexed for 1 min. The supernatant (2 mL) was collected, evaporated under nitrogen stream, and the residue was dissolved in MeOH-H_2O (50:50, *v*/*v*; 1 mL), vortexed for 1 min and filtered prior to high-performance liquid chromatography—fluorescence detection (HPLC-FLD) analysis. A similar protocol was used for the egg sample (5 g), where it was mixed only with 1% CH_3COOH in ACN (10 mL) and PSA was used instead of the C18-zirconia sorbent [38]. Lastly, a fully automated on-line QuEChERS extraction protocol was reported for the extraction of 27 SAs and metabolites from salmon tissue samples. Spiked tissue (1 g) was mixed with 1% CH_3COOH in ACN-H_2O (84:16, *v*/*v*; 5 mL) and the mixture was vortexed at 3200 rpm for 1 min. $MgSO_4$ (1 g) and CH_3COONa (0.1 g) were added and sample extract (2 mL) was aspirated three times, followed by 30 s equilibrium time and centrifuged at 2264× *g* for 5 min. The supernatant (1 mL) was collected and mixed with Z-Sep$^+$ (45 mg), PSA (32 mg), and Na_2SO_4 (0.25 g) and the mixture was vortexed for 1 min and centrifuged at 2264× *g* for 5 min. The supernatant (200 μL) was collected and mixed with MeOH (300 μL) and ammonium formate solution (0.008 M, 500 μL) prior to ultra-high-performance liquid chromatography—electrospray ionization—quadrupole Orbitrap high-resolution mass spectrometry (UHPLC-ESI-Q-Orbitrap-MS) analysis [39].

Two SALLE protocols were also reported. In the first protocol SALLE was reported in combination with magnetic separation for the extraction of eight SAs from fish tissue samples. Spiked tissue (2 g) was mixed with 0.1% formic acid in ACN (5 mL) and the mixture was vortexed for 1 min. NaCl (0.5 g), $MgSO_4$ (2 g) and Fe_3O_4 (100 mg) were added and the mixture was vortexed for 1 min. The magnetic particles were collected by means of a magnet and the supernatant (1 mL) was collected and evaporated to dryness under nitrogen stream at 45 °C. The residue was dissolved in 0.1% formic acid aqueous solution (1 mL), mixed with *n*-hexane (1 mL), and the mixture was centrifuged at 8000 rpm for 3 min. The aqueous phase was collected and passed through a 0.22 μm syringe filter and analyzed with HPLC-MS/MS. The authors emphasized the good recoveries and the reduced time requirements of the developed sample preparation protocol [40]. In the second report, the samples were treated with ethyl acetate and concentrated under vacuum. The analytes were extracted with HCl solution (2 M), the extract was defatted with *n*-hexane, filtered, and mixed with MeOH-ACN-CH_3COONa (5:5:20, *v*/*v*/*v*). Analysis was carried out by HPLC-FLC. However, information could be collected only from the abstract and the further details are not available because the rest of the paper is written in Chinese [41].

2.1.3. SPE

Two SPE protocols were reported for multi-class antibiotic extraction, including 16 SAs, from bovine liver samples [42] and the extraction of SAs from chicken and pork tissue samples [43]. The first extraction protocol employed Oasis HLB cartridges (3 mL, 200 mg) (Waters, Milford, MA, USA). Spiked liver tissue (2 g) was mixed with ACN (10 mL) and EDTA (0.1 M) and the mixture was shaken for 10 min, sonicated for 20 min, and centrifuged at 4000× *g* for 10 min. The supernatant was collected and evaporated under nitrogen stream to 1 mL final volume. H_2O (5 mL) was added and the mixture was vortexed for 15 s and loaded into a SPE cartridge preconditioned with ACN (10 mL) and H_2O (10 mL). The loaded cartridge was washed with H_2O (5 mL) and dried under reduced pressure for 5 min. The analytes were eluted with ACN (10 mL) and the eluate was evaporated under nitrogen stream to 0.5 mL final volume. The reduced eluate was dissolved in mobile phase (400 μL), *n*-hexane (2 mL) was added, and the mixture was vortexed for 30 s and centrifuged at 4000× *g* for 10 min. The mixture was passed through a 0.45 μm filter and analyzed with UHPLC-MS/MS. The developed sample preparation protocol included a SPE clean-up step in order to reduce the liver tissue interferences and utilized SPE cartridges with wide range selectivity in order to achieve multi-class antibiotic extraction [42]. The second extraction protocol employed multi-walled carbon nanotubes (MWCNTs) as the sorbent material. The samples were treated with ACN and the sample extract was dissolved in Na_2HPO_4 buffer (pH 5.5–6.0) and loaded into the SPE cartridges. The cartridges were washed with acetone-hexane (5:95, *v*/*v*) and the analytes were eluted with acetone-dichloromethane (1:1, *v*/*v*). Analysis was carried out by HPLC-UV. However, information could be collected only from the abstract and the further details are not available because the rest of the paper is written in Chinese [43].

Two variations of SPE, a MSPE and a DSPE protocol, were reported for the extraction of SDZ, sulfathiazole, sulfamerazine, SMZ, and SMP from chicken, pork, and shrimp tissue samples [44] and for multi-class antibiotic extraction, including 21 SAs, from animal tissue samples [45], respectively. The MSPE protocol employed a magnetic Fe_3O_4@JUC-48 nanocomposite as the sorbent material. Spiked tissue (2 g) was mixed with ACN (20 mL) and the mixture was vortexed for 5 min, sonicated for 30 min, and kept overnight. The mixture was centrifuged at 10,000 rpm and the supernatant was collected and stored at 4 °C. Sample extract (8 mL) was mixed with Fe_3O_4@JUC-48 (25 mg) and the mixture was vortexed for 8 min. The magnetic sorbent was collected by means of an external magnet and the supernatant was discarded. The analytes were eluted with MeOH-CH_3COOH (95:5, *v*/*v*; 0.8 mL) and sonication for 10 min. The sorbent was separated by means of an external magnet and the eluate was collected, passed through a 0.22 μm nylon filter, and analyzed with HPLC-DAD. Sorbent reusability was studied by applying the sorbent material in several extraction cycles. Between the extractions, the sorbent was washed with MeOH-CH_3COOH (95:5, *v*/*v*; 3 × 1 mL) and MeOH (3 × 1 mL) and dried at 60 °C. The sorbent could be reused for seven extraction cycles without significant adsorption capacity reduction. The authors emphasized the increased sensitivity and the higher recovery values achieved, as well as the reduced extraction time and sorbent requirements of the developed sample preparation protocol [44]. In the DSPE protocol, the samples were treated with Na_2EDTA (0.1 M) and 1% CH_3COOH in ACN, followed by a DSPE clean-up step. Analysis was carried out by HPLC-MS/MS. However, information could be collected only from the abstract and the further details are not available because the rest of the paper is written in Chinese [45].

Other reported approaches included a combined MSPD-homogeneous ionic liquid microextraction (HILME) protocol for the extraction of seven SAs from bovine, chicken, and pork tissue samples [46], an on-line SPME protocol for the extraction of five antimicrobials, including sulfametoxydiazine, sulfamethoxazole, and SQX from chicken and pork tissue and egg samples [47] and a SBSE protocol for the extraction of ten SAs from chicken and pork tissue samples [48]. For the MSPD-HILME protocol, the ionic liquid employed acted both as elution solvent in MSPD and extraction solvent in HILME. Spiked tissue (0.2 g), silica gel (1 g), and 1-butyl-3-methylimidazolium tetrafluoroborate ([C_4MIM][BF_4]) (200 μL) were mixed with a pestle and mortar and the mixture was transferred and placed inside a glass column between absorbent cotton layers. Pure H_2O was passed through the packed mixture and the eluate (3 mL) was mixed with NaCl (0.45 g) and ammonium hexafluorophosphate (2.4 M, 1 mL). The cloudy mixture was centrifuged at 5 °C and 10,000 rpm for 5 min, the supernatant was discarded and the remaining ionic liquid phase was diluted with ACN (300 μL), passed through a 0.22 μm polytetrafluoroethylene membrane filter and analyzed with HPLC-DAD. The authors emphasized the simplicity and the higher extraction recoveries achieved, as well as the reduced reagent requirements of the developed sample preparation protocol [46]. In the on-line SPME protocol, spiked tissue or egg (5 g) was mixed with Na_2SO_4 (5 g) and ACN (10 mL) and the mixture was sonicated for 10 min and centrifuged at 8000 rpm for 5 min. This step was repeated twice and the three sample extracts were combined and evaporated to dryness. The residue was dissolved in

ACN-toluene-*n*-hexane (1:4:45, *v*/*v*/*v*; 25 mL) and used for the on-line SPME. The extraction protocol employed molecularly imprinted monolithic capillary columns with SQX as the template molecule. Activated capillaries were treated with propyltrimethoxysilane and filled with the polymerization mixture that consisted of methacrylic acid (functional monomer), ethylene glycol dimethacrylate (cross-linker), *N*,*N*-dimethylformamide, isooctane, and paraxylene (polymerization and porogenic solvents), while polymerization occurred at 60 °C for 70 h. A prepared monolithic capillary column preplaced the sample loop in an on-line HPLC-UV system and extraction consisted of three steps. In the first step, the sample extract passed through the capillary at a flow rate of 0.15 mL/min so that the analytes came in contact with the imprinted polymeric phase and extracted. In the second step, nitrogen was passed through the capillary so that the residual sample solution was completely removed. Finally, the analytes were eluted from the capillary with 400 µL of mobile phase at a flow rate of 0.15 mL/min. The authors emphasized the increased selectivity and sensitivity, as well as the simplicity and the environmental friendliness of the developed sample preparation protocol [47]. Finally, the SBSE protocol employed poly(vinylphthalimide-*co*-*N*,*N*-methylenebisacrylamide) monolith coated stir bars. Spiked tissue (0.5 g) was mixed with ACN (5 mL), the mixture was sonicated for 15 min and centrifuged at 3000 rpm for 5 min and the supernatant was collected and passed through a 0.45 µm filter. This step was repeated and both supernatants were diluted with Milli-Q H_2O (100 mL). For SBSE, sample solution pH was adjusted (pH 4.0) and NaCl (5%, *w*/*v*) was added, while extraction time was 120 min and liquid desorption time was 60 min. Analysis was carried out by HPLC-MS/MS [48].

All reported literature for the extraction of SAs from animal tissue samples is summarized in Table 2, including recoveries, limit of detection (LOD), limit of quantification (LOQ), or/and decision limit (CC_{α}), decision capability (CC_{β}) values.

Table 2. Extraction of SAs from animal tissue samples.

Food Sample	Analytes	Sample Preparation	Analytical Technique/Run-Time	LOD-LOQ	Recovery (%)	Ref.
bovine tissue	41 antibiotics (15 SAs)	SLE	UHPLC-MS/MS 12 min	CC_α (μg/kg): 104–132 (SAs) CC_β (μg/kg): 108–164 (SAs)	91–109 (SAs)	[31]
fish tissue	41 antibiotics (15 SAs)	SLE	UHPLC-MS/MS 12 min	CC_α (μg/kg): 110.6–126.9 (SAs) CC_β (μg/kg): 121.2–153.7 (SAs)	92–111 (SAs)	[32]
shrimp tissue	SDZ, SMZ, SIX, SDMX and SQX	SLE	HPLC-DAD 40 min	LOD (μg/kg): 15 LOQ (μg/kg): 50	88.6–108.4	[33]
baby foods (combinations of powdered milk, cereal, vegetable, honey and meat)	15 SAs and metabolites	ASE	UHPLC-Orbitrap-MS 10 min	LOD (μg/kg): 0.03–0.17 LOQ (μg/kg): 0.10–0.55	75.5–96.6	[34]
fish tissue	19 antibiotics (9 SAs)	ASE	HPLC-MS/MS N/A	LOD (ng/g): 0.003–0.6	55.2–113 (all analytes)	[35]
chicken, sheep, fish and horse tissue	16 SAs and metabolites	PLE, USE	HPLC-QqLIT-MS/MS 11 min	CC_α (μg/kg): 111.2–161.4 (PLE), 119.3-142.7 (USE) CC_β (μg/kg): 122.4–222.8 (PLE), 138.6-185.5 (USE)	N/A	[36]
bovine, chicken, pork and sheep tissue	22 SAs and metabolites	QuEChERS extraction	HPLC-HRMS 17 min	LOD (μg/kg): 3–26 LOQ (μg/kg): 11–88 CC_α (μg/kg): 101–111 CC_β (μg/kg): 102–122	88–107 (beef muscle)	[37]
chicken tissue and egg	8 SAs	QuEChERS extraction	HPLC-FLD 23 min	LOD (μg/kg): 5.8–19.9 (chicken muscle), 4.1–25.6 (egg) LOQ (μg/kg): 19.2–66.2 (chicken muscle), 13.6–85.4 (egg)	66.9–86.8 (chicken muscle), 65.9–88.1 (egg)	[38]
salmon tissue	27 SAs and metabolites	on-line QuEChERS extraction	UHPLC-ESI-Q-Orbitrap-MS 7 min	CC_α (μg/kg): 0.04–1.34 CC_β (μg/kg): 0.07–2.33	83–109	[39]
fish tissue	8 SAs	SALLE and magnetic separation	HPLC-MS/MS 8 min	LOD (μg/kg): 2.5–10 LOQ (μg/kg): 5–25	74.87–104.74	[40]
shrimp tissue	14 SAs	SALLE	HPLC-FLD N/A	LOD (μg/kg): 1.0–5.0	77.8–103.6	[41]
bovine tissue	39 antibiotics (16 SAs)	SPE	UHPLC-MS/MS 12 min	CC_α (μg/kg): 65–125 (SAs) CC_β (μg/kg): 81–150 (SAs)	85–110 (SAs)	[42]
chicken and pork tissue	SAs	SPE	HPLC-UV N/A	LOD (mg/L): 0.003 LOQ (mg/L): 0.01	> 70	[43]
chicken, pork and shrimp tissue	SDZ, sulfathiazole, sulfamerazine, SMZ and SMP	MSPE	HPLC-DAD 16 min	LOD (ng/g): 1.73–5.23 LOQ (ng/g): 3.97–15.89	81.4–101.3 (chicken), 76.1–102.6 (pork), 79.2–102.5 (shrimp)	[44]
animal tissue	63 veterinary drugs (21 SAs)	DSPE	HPLC-MS/MS N/A	LOD (μg/kg): 0.1–3.0 (all analytes) LOQ (μg/kg): 0.5–10.0 (all analytes)	62.2–112.0 (all analytes)	[45]
bovine, chicken and pork tissue	7 SAs	MSPD-HILME	HPLC-DAD 30 min	LOD (μg/kg): 4.3–13.4 LOQ (μg/kg): 14.2–44.8	85.4–95.1 (kidney), 104.5–118.0 (liver), 85.5–112.3 (muscle)	[46]
chicken and pork tissue and egg	5 antimicrobials (sulfametoxydiazine, sulfamethoxazole and SQX)	on-line SPME	HPLC-UV 20 min	LOD (μg/L): 0.10–0.14 (SAs) LOQ (μg/L): 0.39–0.47 (SAs)	84.1–99.6 (SAs in chicken), 87.0–108.2 (SAs in pork), 80.1–101.4 (SAs in egg)	[47]
chicken and pork tissue	10 SAs	SBSE	HPLC-MS/MS N/A	LOD (μg/kg): 0.0012–0.0061 (pork), 0.0021–0.0146 (chicken) LOQ (μg/kg): 0.0040–0.0203 (pork), 0.0066–0.0487 (chicken)	62.4–109.9 (pork), 55.2–109.1 (chicken)	[48]

2.2. Milk Samples

2.2.1. LLE

An LLE protocol was reported for the extraction of nine SAs from milk samples. Spiked milk (100 µL) was mixed with acidified dichloromethane (800 µL) and the mixture was sonicated for 10 min and centrifuged at 93,000× *g* for 10 min. The organic phase was collected and the step was repeated. The combined organic phases were evaporated to dryness under nitrogen stream at 40 °C and the residue was dissolved in acidified MeOH (100 µL) and filtered prior to HPLC-MS/MS analysis. The authors emphasized the reduced sample and reagent requirements of the developed sample preparation protocol, while a SPE clean-up step was omitted in order to simplify and reduce the cost of the protocol [49]. DLLME and modified QuEChERS extraction, were reported for the extraction of nine SAs from milk samples. In both cases off-line derivatization was conducted with fluorescamine (50 µL) and sonication for 15 min prior to HPLC-FLD analysis. For the DLLME protocol, spiked milk (30 mL) was mixed with 20% trichloroacetic acid aqueous solution (15 mL) and the mixture was vortexed for 10 s and centrifuged at 6000 rpm for 5 min. The supernatant was passed through a 0.2 µm filter and pH was adjusted to pH 4.0–4.5. Treated supernatant (5 mL) was mixed with chloroform (1000 µL, extraction solvent) and ACN (1900 µL, dispersive solvent) and the mixture was shaken until a cloudy solution was formed and centrifuged at 6000 rpm for 5 min. The chloroform phase was collected with a syringe, evaporated under nitrogen stream and the residue was dissolved in Tris buffer (pH 7.0, 1.5 mL) and passed through a 0.2 µm nylon filter. For the QuEChERS protocol, spiked milk (2 mL) was mixed with H_2O (8 mL), vortexed for 10 s, 5% CH_3COOH in ACN (10 mL) was added, and the mixture was shaken for 30 s. The QuEChERS (C18, $MgSO_4$ and PSA) were added and the mixture was shaken for 1 min and centrifuged at 4000 rpm for 5 min. The supernatant (1.5 mL) was collected, evaporated under nitrogen stream and the residue was dissolved in Tris buffer (pH 7.0, 1.5 mL) and passed through a 0.2 µm nylon filter. Both developed sample preparation protocols were simple, fast, and environmentally friendly, providing good recoveries. When compared, DLLME gave lower LOD values and higher recoveries, while QuEChERS extraction was more reproducible with higher throughput [50].

2.2.2. Salting-Out Extraction

SALLE, combined with SPE [51] and a miniaturized SALLE protocol [52], were reported for multi-veterinary drug extraction, including 26 SAs, from milk samples and sulfonamide from tea beverage, water, milk, honey, plasma, blood, and urine samples, respectively. The first protocol employed Oasis HLB Plus cartridges (225 mg) (Waters, Milford, MA, USA). For the SALLE step, spiked milk (5 g) was mixed with oxalic acid-EDTA buffer (pH 3.0, 5 mL) and ACN (10 mL) and the mixture was shaken for 30 s and centrifuged at 3000 rpm for 5 min. The supernatant was collected, $(NH_4)_2SO_4$ (1 g) was added and the mixture was shaken for 2 min, left for 2 min and centrifuged at 3000 rpm for 3 min. Three layers resulted after centrifuging, and the upper ACN and the lower aqueous layer were used for the SPE clean-up step, while the middle layer was the milk fat. The SPE cartridges were preconditioned with MeOH (10 mL), H_2O (10 mL), and oxalic acid-EDTA buffer (pH 3.0, 2 mL), loaded with the aqueous layer, washed with buffer (2 mL) and the analytes were eluted with the ACN layer (10 mL) and MeOH (5 mL). The eluate (3 ML) was evaporated to 0.1–0.2 mL final volume under nitrogen stream at 50 °C for 20 min and the residue was dissolved in CH_3COONH_4 solution (0.1 M, 1 mL), vortexed for 30 s, and passed through a 0.45 µm filter device prior to UHPLC-ESI-Q-Orbitrap-MS analysis [51]. The second protocol employed two 1 mL syringes coupled via their tips that contained the sample solution and extraction solution, respectively. For the preparation of milk samples, spiked milk (1 mL) was mixed with ACN-MeOH-H_2O (40:20:20, *v*/*v*/*v*; 1 mL), the mixture was centrifuged at 3000 rpm for 10 min. For the preparation of honey samples, spiked honey was diluted with H_2O at a concentration of 0.1 g/mL and the mixture was homogenized and centrifuged at 4000 rpm for 20 min. Milk or honey supernatant (0.5 mL) was retracted with the sample syringe A, NaCl (250 mg/mL) was added, and the mixture was vortexed for 20 s and adjusted to pH 7.0 with NaOH solution (0.1 M). ACN (250 mL) was retracted with the extraction solution syringe B and both syringes were coupled and held vertically. The extraction was conducted by injecting the extraction solvent from syringe B into syringe A, forming a cloudy solution and the content was pumped back to syringe B. This step was repeated five times and the mixture was left in syringe B vertically for 2 min. After phase formation, the upper phase was collected and injected for HPLC-UV analysis. The authors emphasized the reduced organic solvent and sample requirements, as well as the simplicity and the improved extraction efficiency of the developed sample preparation protocol [52].

Two ATPS extraction protocols were reported in the literature for the extraction of SAs from milk. A modified ATPS protocol was reported for the extraction of SDZ and SMZ from milk, egg, and water samples. The extraction protocol employed polyoxyethylene lauryl ether and $Na_2C_4H_4O_6$ in order to form the polymer-organic salt extraction system. For the preparation of milk and egg samples, spiked milk (50 mL) or homogenized spiked egg (50 mL) was mixed with 10% trichloroacetic acid solution (20 mL) and H_2O to 100 mL final volume and the mixture was shaken and centrifuged at 4000 rpm for 30 min. The supernatant was collected and passed through a 0.45 μm filter in order to remove the proteins. The filtrate was added to a mixture of polyoxyethylene lauryl ether (0.027 g/mL) and $Na_2C_4H_4O_6$ (0.180 g/mL) and filled with H_2O to 10 mL final volume. The mixture was placed in a heated water bath under continuous stirring for 20 min and after the phase formation, the upper phase was collected and analyzed with HPLC-UV [53]. An ionic liquid ATPS extraction protocol was reported for the extraction of six SAs from milk samples. Spiked milk (5 mL) diluted with pure H_2O (2.5 mL) was mixed with 10% $HClO_4$ aqueous solution (500 μL) and the mixture was shaken for 2 min and centrifuged at 10,000 rpm for 10 min. The supernatant was collected and passed through a 0.45 μm filter. Butyl-3-methylimidazolium tetrafluoroborate ([C_4MIM][BF_4]) (300 μL) and $C_6H_5Na_3O_7 \cdot 2H_2O$ (3 g) were added and the mixture was shaken and centrifuged at 10,000 rpm for 10 min. The upper IL phase was collected, diluted with ACN at 1:1 ratio, sonicated, and passed through a 0.22 μm filter. Analysis was carried out by HPLC-UV. The authors emphasized the reduced organic solvent consumption of the developed sample preparation protocol in comparison with classic LLE, as well as the combination of sample preconcentration and clean-up in one step [54].

2.2.3. SPE

Two SPE protocols were reported for multi-veterinary drug extraction, including 18 SAs [55], and the extraction of six SAs [56] from milk samples. The first extraction protocol employed Oasis MCX cartridges (3 mL, 60 mg) (Waters, Milford, MA, USA). Spiked milk (2 g) was mixed with 1% CH_3COOH in ACN (5 mL) and the mixture was vortexed for 30 s and centrifuged at 5000 rpm for 12 min. The supernatant was evaporated to dryness under nitrogen stream at 40 °C and the residue was dissolved in HCl solution (0.1 M, 3 mL) and loaded to a SPE cartridge, conditioned with MeOH (3 mL) and HCl solution (0.1 M, 3 mL). The loaded cartridge was washed with HCl solution (0.1 M, 3 mL) and MeOH (3 mL) and the analytes were eluted with 10% ammonia in ACN (4 mL). The eluate was evaporated to dryness under nitrogen stream at 40 °C and the residue was dissolved in 0.1% CH_3COOH in CH_3COONH_4-MeOH (90:10, *v*/*v*; 1 mL) and passed through a 0.22 μm filter prior to ultra-high-performance liquid chromatography—electrospray ionization—mass spectrometry (UHPLC-ESI-MS/MS) analysis [55]. The second extraction protocol employed multi-template MIPs prepared by sol-gel synthesis. Spiked milk (1 g), deproteinized with ACN, was loaded into MISPE cartridges packed with the prepared MIPs (30 mg) and left for 15 min to equilibrate. The loaded cartridge was washed with MeOH (2 mL) and the analytes were eluted with 1% CH_3COOH-MeOH-ACN (50:10:40, *v*/*v*/*v*; 2 mL) at a flow rate of 1 mL/min and the eluate was evaporated to dryness under nitrogen stream. Analysis was carried out by HPLC-DAD. MISPE cartridge conditioning was omitted as an unnecessary step, thus reducing the time of the developed sample preparation protocol [56].

Three MSPE protocols were reported for the extraction of nine SAs [57], SMP, SMZ, sulfamethoxazole and sulfachloropyridazine [58], and five SAs [59] from milk samples. The first protocol employed silica-based magnetic sorbent material. Magnetic sorbent (0.1 g), conditioned with MeOH (5 mL) and sonication for 5 min and washed with deionized H_2O (2 × 10 mL), was added to spiked milk (10 mL). The mixture was sonicated for 15 min, the magnetic sorbent was collected by means of a magnet and the supernatant was discarded. The sorbent was washed with acetate buffer (pH 4.0, 3 × 5 mL) and the analytes were with 10^{-3} M NaOH in MeOH (5 mL) for 5 min. The eluate was collected, evaporated to dryness under nitrogen stream and the residue was dissolved in 1% formic acid aqueous solution (500 μL) and passed through a 0.2 μm nylon filter prior to HPLC-DAD analysis. The authors emphasized the simplicity, higher recovery values, and the lower organic solvent requirements of the developed sample preparation protocol in comparison with classic SPE [57]. The second protocol employed a magnetic hyper cross-linked polystyrene composite as the sorbent material. Spiked milk (25 mL) was agitated for 15 min and the magnetic sorbent was added (20 mg). The sample pH was adjusted to pH 5.0, extraction/stirring time was 10 min and analytes were eluted with ACN (2 × 1 mL) and sonication for 5 min. Analysis was carried out by high-performance liquid chromatography—amperometric detection (HPLC-AD). The proposed magnetic composite combined large surface area, high adsorption, and magnetic separation, while a small amount could be used for the extraction from large volumes of untreated milk. The authors emphasized the good recovery's simplicity of the sample preparation protocol, as well as the reduced time and solvent requirements in comparison with classic sample preparation techniques [58]. The last MSPE protocol employed a magnetic graphene-based composite ($CoFe_2O_4$-graphene) as the sorbent material. Spiked milk (1.5 mL) was mixed with 15% $HClO_4$ aqueous solution (0.2 mL) and the mixture was vortexed for 30 s and centrifuged at 14,000 rpm for 5 min. The supernatant was collected, diluted with deionized H_2O (100 mL), and the pH was adjusted to pH 4.0 and mixed with the magnetic sorbent (15 mg). The magnetic sorbent was previously conditioned with MeOH (5 mL), H_2O (5 mL) and sonication for 5 min. The mixture was shaken for 20 min and vortexed for 2 min. The magnetic particles were collected by means of a magnet and the supernatant was discarded. The analytes were eluted with 5% CH_3COOH in MeOH (0.5 mL) and the eluate was passed thought a 0.22 μm filter prior to HPLC-UV analysis. The authors emphasized the simplicity, low LOD values, and improved recoveries of the developed sample preparation protocol in comparison with other reported methods. The magnetic sorbent displayed increased extraction efficiency for the analytes, while it could be reused after washing with ACN and ultrapure H_2O [59].

2.2.4. Other Extraction Techniques

Interesting approaches reported for the extraction of SAs from milk samples include FPSE, graphene-modified melamine sponge (GMeS) microextraction and miniaturized syringe assisted extraction (mini-SAE). The FPSE protocol was reported for the extraction of SMZ, SIX, and SDMX from milk samples. The extraction protocol employed highly polar sol-gel poly(ethylene glycol) coated cotton cellulose fabric segments as the sorbent material. FPSE media incubated in MeOH-ACN (50:50, *v*/*v*; 2 mL) for 5 min and rinsed with H_2O (2 mL), was introduced into spiked whole milk (1 g) for 30 min. The fabric-milk system was stirred by means of a magnetic stirrer for 30 min and the extraction media was transferred and incubated in MeOH (250 μL) for 8 min and ACN (250 μL) for 5 min. The extract was filtered prior to HPLC-UV analysis. The coated fabric was washed with ACN-MeOH (50:50, *v*/*v*; 2 mL) for 5 min, left to dry for 5–10 min, and kept in an air-tight container between extractions and could be reused for up to 30 times. The developed sample preparation protocol eliminated deproteinization and evaporation/reconstitution, thus reducing the extraction time and the errors resulting from these steps. The proposed fabric sorbent could be applied directly into the untreated milk sample offering a simpler extraction protocol and higher recoveries. Furthermore, the fabric sorbent displayed high chemical and solvent stability that allows the use of the suitable extraction solvent for sample analysis with multiple chromatographic techniques [60]. The GMeS microextraction protocol was reported for the extraction of eight SAs from milk, egg, and water. The extraction protocol employed novel graphene-modified melamine sponges as the sorbent material. For the preparation of milk samples, spiked milk (15 mL) was defatted with centrifuging at 4000 rpm and 4 °C for 10 min and deproteinized with 15% trichloroacetic acid solution (1 mL for every 10 mL defatted sample solution), vortexing for 1 min, and centrifuging at 4000 rpm for 5 min. The supernatant was collected and mixed with NaCl (6% *w*/*v*) and centrifuged. The supernatant was used for the GMeS extraction. For the preparation of egg samples, homogenized spiked egg (1 g) was mixed with double distilled H_2O (8.7 mL), 15% trichloroacetic acid solution (0.3 mL), and NaCl (6% *w*/*v*) and stirred for 1 min prior to extraction. GMeS cubes conditioned with MeOH and distilled H_2O were placed inside the sample solution (10 mL) and stirred at 600 rpm for 30 min. The cube was collected, placed into a syringe cartridge, rinsed with H_2O, and squeezed in order to remove the absorbed sample. The analytes were eluted with 5% ammonia in ACN (2 × 1 mL), the eluate was evaporated to dryness under nitrogen stream and the residue was dissolved in H_2O-ACN (70:30, *v*/*v*; 100 µL) and sonicated for 1 min prior to HPLC-DAD analysis. The authors emphasized the simple and rapid preparation and easier handling, as well as the improved recoveries and environmental friendliness of developed GMeS material in comparison with other sorbents found in the literature [61]. The mini-SAE protocol was reported for the extraction of SDZ and sulfamonomethoxine from milk samples. The extraction protocol employed a poly (hydroxyethyl methacrylate) polymer as the sorbent material. Spiked milk (50 g) was mixed with 16% lead acetate aqueous solution (3 mL) and the mixture was stirred for 5 min and centrifuged at 4000 rpm for 4 min. The supernatant was collected, 16% lead acetate aqueous solution (2 mL) was added, and the mixture was centrifuged at 4000 rpm for 4 min. The supernatant (1 mL) was loaded into the mini-SAE device packed with the polymer sorbent (50 mg) and conditioned with MeOH (2 mL) and H_2O (2 mL). The device was washed with H_2O (1 mL) and the analytes were eluted with 5% CH_3COOH in MeOH (3 mL). The eluate was evaporated to dryness under nitrogen stream and the residue was dissolved in phosphate buffer (pH 4.0, 1 mL) and derivatized with fluorescamine prior to HPLC-FLD analysis [62].

2.3. Milk Product Samples

Three protocols were also reported for the extraction of SAs from milk products (baby formula, cheese, and butter). Firstly, a SLE protocol was reported for multi-veterinary drug extraction, including 24 SAs, from baby formula samples. Spiked formula (1 g) was mixed with EDTA aqueous solution (0.05 M, 10 mL), the mixture was vortexed, 0.1% formic acid in ACN (10 mL) was added, and the mixture was vortexed, shaken for 15 min and centrifuged at 2000 rcf for 10 min. The supernatant (2 mL) was collected, evaporated to dryness under nitrogen stream at 40 °C, and the residue was dissolved in H_2O-ACN (75:25, *v*/*v*; 1 mL). Analysis was carried out by UHPLC-MS/MS. A clean-up step was omitted due to low recoveries for β-lactams, tetracyclines and dyes, and variable recoveries for the other analytes [63]. A QuEChERS extraction protocol was reported for multi-veterinary drug extraction, including sulfachloropyridazine, sulfadimidine, SDMX, and SQX, from cheese samples. Spiked cheese (10 g) was mixed with 1% CH_3COOH in ACN (10 mL) and Na_2EDTA solution (0.1 M, 10 mL) and the mixture was vortexed for 1 min. $MgSO_4$ (4 g) and CH_3COONa (1 g) were added and the mixture was stirred for 1 min and centrifuged at 4500 *g* for 5 min. The supernatant (2 mL) was collected and passed through a 0.2 µm nylon filter and the filtrate (1 mL) was diluted with 0.01% formic acid solution-MeOH (50:50, *v*/*v*; 1 mL) prior to analysis with UHPLC-MS/MS. The developed sample preparation protocol enabled the extraction of multiple veterinary drugs, in comparison with other similar protocols, that were used for the extraction of a single antibiotic or antibiotic group [64]. An ionic liquid—magnetic bar—liquid-phase microextraction (IL-MB-LPME) was reported for the extraction of eight SAs from butter samples. The extraction protocol employed magnetic hollow fibers as the extraction configuration and 1-octyl-3-methylimidazolium hexafluorophosphate ([C_8MIM][PF_6]) immobilized on the hollow fiber micropores as the extraction solvent. Spiked butter (30 g) was added into a vessel containing eight magnetic fibers and Na_2SO_4 aqueous solution (3 M, 6 mL) and the vessel was sealed and placed into a water bath at 45 °C and magnetic stirring at 500 rpm for 25 min. The magnetic fibers were collected by means of a magnet, washed with hexane (1 mL), and the analytes were eluted with MeOH (200 μL) and sonication for 3 min. The eluate was collected, mixed with Na_2SO_4 (100 mg) and the supernatant was passed through a 0.22 μm filter. Analysis was carried out by HPLC-UV. The Na_2SO_4 aqueous solution acted both as the extraction solvent for the extraction of the SAs from the butter and as the sample solution for the IL magnetic hollow fibers, thus the developed sample preparation protocol combined analyte extraction, clean-up, and preconcentration in one step [65].

2.4. *Egg Samples*

Two SPE protocols were reported for the extraction of SAs from egg samples. The first SPE protocol was reported for the extraction of 13 SAs from egg samples. The extraction protocol employed Strata-X SCX cartridges (Phenomenex, Macclesfield, UK). Homogenized spiked egg (10 g) was adjusted to pH 5.0–6.0 with 10% CH_3COOH solution (900 μL) for 15 min. Chloroform-acetone (50:50, *v*/*v*; 30 mL) was added and the mixture was shaken for 10 min and sonicated for 20 min. NaCl (3 g) and Na_2SO_4 (3 g) were added and the mixture was centrifuged at 2209 *g* and 10 °C for 10 min and placed at −70 °C for 30 min. The organic phase was collected (25 mL), mixed with CH_3COOH (2.5 mL), and loaded into the SPE cartridge conditioned with *n*-hexane (2 × 3 mL) and acetone-5% CH_3COOH in chloroform (50:50, *v*/*v*; 2 × 3 mL). The loaded cartridge was washed with H_2O (5 mL) and MeOH (5 mL) and the analytes were eluted with MeOH-ammonia solution (97.5:2.5, *v*/*v*; 13 mL). The eluate was evaporated to dryness under nitrogen stream at 45 °C and the residue was dissolved in mobile phase (0.5 mL), mixed with *n*-hexane (0.5 mL), and centrifuged at 2209× *g* and 20 °C for 10 min. The lower phase was collected, centrifuged for another 10 min, and the supernatant was analyzed with HPLC-DAD [66]. The second SPE protocol was reported for the extraction of SDZ from egg samples. The extraction protocol employed SDZ imprinted microspheres (100 mg) packed into a glass syringe conditioned with MeOH (5 mL) and Milli-Q H_2O (5 mL). Spiked egg yolk (2 g) and white (2 g) were respectively mixed with MeOH (10 mL) and the mixture was sonicated for 10 min. The supernatants were collected and the step was repeated for both egg yolk and white. All collected supernatants were combined, concentrated to 10 mL final volume, and loaded to the MISPE cartridge. The loaded cartridge was washed with MeOH-H_2O (30:70, *v*/*v*; 1 mL) and the analyte was eluted with MeOH (1 mL). The eluate was analyzed directly with HPLC-DAD. The authors emphasized the clean-up efficiency and analyte preconcentration achieved by the developed extraction protocol, while sample defatting was not necessary [67]. Additionally, QuEChERS extraction [38], on-line SPME [47], ATPS [53], and GMeS microextraction [61] were reported for the extraction of SAs from egg samples and sample preparation protocols are given in detail in Sections 2.1 and 2.2.

2.5. *Honey Samples*

An on-line SPE protocol was reported for the extraction of 15 SAs from honey samples. The extraction was achieved in a Zorbax Extended C-18 (12 mm × 4.6 mm; 5 μm) column (Agilent, Santa Clata, CA, USA). Spiked honey (1 g) was hydrolyzed with HCl solution (3 M, 800 μL) for 90 min and neutralized with citrate buffer (pH 3.5, 200 μL) and NaOH solution (10 M, 240 μL). The SAs were derivatized with 0.2% fluorescamine (200 μL) and the sample solution was passed through a 0.22 μm filter and injected to the on-line SPE-HPLC-FLD system. The authors emphasized the reduced organic solvent and sample requirements, as well as the simplicity, environmental friendliness, and increased selectivity and sensitivity of the automated SPE protocol [68]. Additionally, a miniaturized SALLE [52] protocol was reported for the extraction of sulfonamide from honey samples and details are provided in Section 2.2.

All reported literature for the extraction of SAs from milk and milk product, egg, and honey samples is summarized in Table 3.

Table 3. Extraction of SAs from milk and milk product, egg, and honey samples.

Food Sample	Analytes	Sample Preparation	Analytical Technique/Run-Time	LOD-LOQ	Recovery (%)	Ref.
milk	9 SAs	LLE	HPLC-MS/MS 30 min	LOQ (μg/kg): 12.5–45 CC_α (μg/kg): 106–122 CC_β (μg/kg): 112–145	89–105	[49]
milk	9 SAs	DLLME, QuEChERS extraction	HPLC-FLD 15 min	LOD (μg/L): 0.60–1.21 (DLLME), 1.15–2.73 (QuEChERS) LOQ (μg/L): 2.01–4.02 (DLLME), 3.85–9.09 (QuEChERS)	90.8–104.7 (DLLME), 83.6–104.8 (QuEChERS)	[50]
milk	105 veterinary drugs (26 SAs)	SALLE and SPE	UHPLC-ESI-Q-Orbitrap-MS 14 min	LOQ (μg/kg): 1.0 (all analytes)	71–120 (all analytes)	[51]
tea beverage, water, milk, honey, plasma, blood and urine	sulfonamide	miniaturized SALLE	HPLC-UV 7 min	LOD (ng/mL): 0.3 LOQ (ng/mL): 1.0	96.66 (tea beverage), 76.67 (milk), 43.33 (honey)	[52]
milk, egg and water	SDZ and SMZ	ATPS extraction	HPLC-UV N/A	LOD (pg/mL): 2.92–3.64 (milk), 2.90–3.49 (egg) LOQ (pg/mL): 9.73–12.15 (milk), 9.66–11.62 (egg)	97.14–99.52 (milk), 96.90–99.30 (egg)	[53]
milk	6 SAs	ATPS extraction	HPLC-UV 30 min	LOD (ng/mL): 2.04–2.84 LOQ (ng/mL): 6.73–9.37	72.32–108.96	[54]
milk	38 veterinary drugs (18 SAs)	SPE	UHPLC-ESI-MS/MS 13 min	CC_α (μg/kg): 109–114 (SAs) CC_β (μg/kg): 116–123 (SAs)	87–119 (all analytes)	[55]
milk	6 SAs	SPE	HPLC-DAD 15.3 min	LOD (μg/kg): 1.9–13.3 LOQ (μg/kg): 5.6–42.2 CC_α (μg/kg): 101.9–113.5 CC_β (μg/kg): 114.4–135.4	N/A	[56]
milk	9 SAs	MSPE	HPLC-DAD 35 min	LOD (μg/L): 7–14 CC_α (μg/kg): 108.86–117.16 CC_β (μg/kg): 117.73–134.32	81.88–114.98	[57]
milk and water	SMP, SMZ, sulfamethoxazole and sulfachloropyridazine	MSPE	HPLC-AD N/A	LOD (ng/mL): 2.0–2.5 (milk) LOQ (ng/mL): 6.0–7.5 (milk)	92–105 (milk)	[58]
milk	5 SAs	MSPE	HPLC-UV 8 min	LOD (μg/L): 1.16–1.59 LOQ (μg/L): 3.52–4.81	62.0–104.3	[59]
milk	SMZ, SIX and SDMX	FPSE	HPLC-UV 6.5 min	CC_α (μg/kg): 114.4–116.5 CC_β (μg/kg): 104.1–118.5	93–107	[60]
milk, egg and water	8 SAs	GMeS microextraction	HPLC-DAD 30 min	LOQ (μg/kg): 0.31–0.91 (milk), 0.96–1.32 (egg)	90–105 (milk), 90–108 (egg)	[61]

Table 3. *Cont.*

Food Sample	Analytes	Sample Preparation	Analytical Technique/Run-Time	LOD-LOQ	Recovery (%)	Ref.
milk	SDZ and sulfamonomethoxine	mini-SAE	HPLC-FLD N/A	LOD (ng/g): 0.19–0.26 LOQ (ng/g): 0.67–0.87	85.6–100.3	[62]
baby formula	150 veterinary drugs (24 SAs)	SLE	UHPLC-MS/MS 17.5 min	LOQ (ng/g): 1–10 (all analytes)	50–120 (all analytes)	[63]
cheese	17 veterinary drugs (sulfachloropyridazine, sulfadimidine, SDMX and SQX)	QuEChERS extraction	UHPLC-MS/MS 8.5 min	LOD (μg/kg): 0.2–1.7 (SAs) LOQ (μg/kg): 0.7–5.5 (SAs) CC_{α} (μg/kg): 3.4–5.8 (SAs) CC_{β} (μg/kg): 5.7–10.2 (SAs)	72.5–106.3 (SAs)	[64]
butter	8 SAs	IL-MB-LPME	HPLC-UV 30 min	LOD (μg/kg): 1.20–2.17 LOQ (μg/kg): 4.00–7.25	73.25–103.85	[65]
egg	13 SAs	SPE	HPLC-DAD 45 min	LOD (μg/kg): 0.30–1.29 LOQ (μg/kg): 0.92–3.92 CC_{α} (μg/kg): 11.3–18.5 CC_{β} (μg/kg): 13.2–27.3	45.2–87.5	[66]
egg	SDZ	SPE	HPLC-DAD N/A	LOD (μg/L): 0.06 (egg yolk), 0.05 (egg white) LOQ (μg/L): 0.20 (egg yolk), 0.17 (egg white)	78.22–86.10	[67]
honey	15 SAs	on-line SPE	HPLC-FLD 30 min	LOD (ng/g): 0.1–1.0	76–108	[68]

3. Conclusions

The most recent literature regarding the extraction of SAs from food samples was successfully reported. In the case of animal tissue samples, SLE [31–33], ASE [34–36], QuEChERS extraction [37,38], SALLE [40,41], and SPE [42] were the most reported methodologies for the extraction of SAs, along with other antibiotic or veterinary drugs, followed by reports of USE [36], DSPE [45], and SBSE [48]. Reports of on-line techniques include a fully automated on-line QuEChERS extraction protocol [39] and an on-line SPME protocol that utilized molecularly imprinted monolithic capillary columns [47]. Reported novel solid-phase sorbent materials include MWCNTs utilized in a SPE protocol [43], and Fe_3O_4@JUC-48 nanocomposite utilized in a MSPE protocol [44]. An ionic liquid application was also reported in a MSPD-HILME protocol [46]. SPE [55], MSPE [57], and ATPS extraction [53,54] were the most reported approaches for the extraction of SAs from milk samples. Other approaches include LLE [49], DLLME [50], as well as a modified QuEChERS extraction protocol that employed C18, $MgSO_4$ and PSA [50], SALLE combined with SPE [51], and a miniaturized SALLE protocol [52]. Interesting approaches include FPSE [60], GMeS microextraction [61], and mini-SAE [62]. Reported novel materials include a magnetic hyper cross-linked polystyrene composite [58] and a magnetic graphene-based composite [59] utilized in MSPE protocols and multi-template MIPs prepared by sol-gel synthesis utilized in a SPE protocol [56]. Furthermore, SLE [63], QuEChERS extraction [64], and IL-MB-LPME [65] were reported for the extraction of SAs from milk products. In the case of egg samples, SPE [66,67] was the main reported technique, while QuEChERS extraction [38], on-line SPME [47], ATPS [53], and GMeS microextraction [61] protocols included eggs, along with other food matrices. Finally, for honey samples only an on-line SPE [68] and a miniaturized SALLE [52] protocol were reported in the recent literature.

Author Contributions: The authors have equally contributed to the manuscript.

Funding: This research received no external funding.

Conflicts of Interest: The authors declare no conflict of interest.

References

1. The Council of The European Union. Council Directive 96/23/EC. *Off. J.* **1996**, *125*, 10–32.
2. PAGE, S.W.; GAUTIER, P. Use of antimicrobial agents in livestock. *Rev. Sci. Tech. OIE* **2012**, *31*, 145–188. [CrossRef]
3. Moreno, L.; Lanusse, C. Specific Veterinary Drug Residues of Concern in Meat Production. In *New Aspects of Meat Quality*; Elsevier: New York, NY, USA, 2017; pp. 605–627, ISBN 9780081005934.
4. Croubels, S.; Daeseleire, E. *Veterinary Drug Residues in Foods*; Woodhead Publishing Limited: Sawston, UK, 2012; ISBN 9780857090584.
5. Moreno, L.; Lanusse, C. Veterinary Drug Residues in Meat-Related Edible Tissues. In *New Aspects of Meat Quality*; Elsevier: New York, NY, USA, 2017; pp. 581–603, ISBN 9780081005934.
6. Baran, W.; Adamek, E.; Ziemiańska, J.; Sobczak, A. Effects of the presence of sulfonamides in the environment and their influence on human health. *J. Hazard. Mater.* **2011**, *196*, 1–15. [CrossRef] [PubMed]
7. EUR-Lex. Available online: http://eur-lex.europa.eu/legal-content/EN (accessed on 14 January 2017).
8. Prescott, J.F. Sulfonamides, Diaminopyrimidines, and Their Combinations. In *Antimicrobial Therapy in Veterinary Medicine*; John Wiley & Sons, Inc: Hoboken, NJ, USA, 2013; pp. 279–294.
9. Reybroeck, W.; Daeseleire, E.; De Brabander, H.F.; Herman, L. Antimicrobials in beekeeping. *Vet. Microbiol.* **2012**, *158*, 1–11. [CrossRef] [PubMed]
10. Maximum Residue Limits, Codex Alimentarius. Available online: http://www.fao.org/fao-who-codexalimentarius/codex-texts/maximum-residue-limits/pt/ (accessed on 15 May 2018).
11. U.S. Food & Drug Administration CFR, Code of Federal Regulations Title 21. Available online: https://www.accessdata.fda.gov/scripts/cdrh/cfdocs/cfcfr/cfrsearch.cfm?cfrpart=556 (accessed on 15 May 2018).
12. Otles, S.; Ozyurt, V.H. Sampling and Sample Preparation. In *Handbook of Food Chemistry*; Springer: Berlin/Heidelberg, Germnay, 2015; pp. 151–164.

13. Saha, S.; Singh, A.K.; Keshari, A.K.; Raj, V.; Rai, A.; Maity, S. Modern Extraction Techniques for Drugs and Medicinal Agents. In *Ingredients Extraction by Physicochemical Methods in Food*; Elsevier: New York, NY, USA, 2018; pp. 65–106, ISBN 9780128115213.
14. Pérez-Rodríguez, M.; Pellerano, R.G.; Pezza, L.; Pezza, H.R. An overview of the main foodstuff sample preparation technologies for tetracycline residue determination. *Talanta* **2018**, *182*, 1–21. [CrossRef] [PubMed]
15. Tang, Y.Q.; Weng, N. Salting-out assisted liquid-liquid extraction for bioanalysis. *Bioanalysis* **2013**, *5*, 1583–1598. [CrossRef] [PubMed]
16. Iqbal, M.; Tao, Y.; Xie, S.; Zhu, Y.; Chen, D.; Wang, X.; Huang, L.; Peng, D.; Sattar, A.; Shabbir, M.A.B.; et al. Aqueous two-phase system (ATPS): An overview and advances in its applications. *Biol. Proc. Online* **2016**, *18*, 18. [CrossRef] [PubMed]
17. Maciel, E.V.S.; de Toffoli, A.L.; Lanças, F.M. Recent trends in sorption-based sample preparation and liquid chromatography techniques for food analysis. *Electrophoresis* **2018**, 1–48. [CrossRef] [PubMed]
18. Bitas, D.; Samanidou, V. Molecularly imprinted polymers as extracting media for the chromatographic determination of antibiotics in milk. *Molecules* **2018**, *23*, 316. [CrossRef] [PubMed]
19. Vasconcelos, I.; Fernandes, C. Magnetic solid phase extraction for determination of drugs in biological matrices. *TrAC Trends Anal. Chem.* **2017**, *89*, 41–52. [CrossRef]
20. Kazantzi, V.; Anthemidis, A. Fabric Sol–gel Phase Sorptive Extraction Technique: A Review. *Separations* **2017**, *4*, 20. [CrossRef]
21. Cobos, Á.; Díaz, O. Chemical Composition of Meat and Meat Products. In *Handbook of Food Chemistry*; Springer: Berlin/Heidelberg, Germnay, 2015; pp. 471–510.
22. Coppes Petricorena, Z. Chemical Composition of Fish and Fishery Products. In *Handbook of Food Chemistry*; Springer: Berlin/Heidelberg, Germnay, 2015; pp. 403–435.
23. Mehta, B.M. Chemical Composition of Milk and Milk Products. In *Handbook of Food Chemistry*; Springer: Berlin/Heidelberg, Germnay, 2015; pp. 511–553.
24. Sunwoo, H.H.; Gujral, N. Chemical Composition of Eggs and Egg Products. In *Handbook of Food Chemistry*; Springer: Berlin/Heidelberg, Germnay, 2015; pp. 331–363.
25. U.S. Food and Drug Administration. Laboratory Methods, Drug & Chemical Residues Methods. Available online: https://www.fda.gov/Food/FoodScienceResearch/LaboratoryMethods/ucm2006950.htm (accessed on 17 May 2018).
26. U.S. Food and Drug Administration. Laboratory Methods, Analytical Methods for Residues of Chloramphenicol & Related Compounds in Foods. Available online: https://www.fda.gov/Food/FoodScienceResearch/LaboratoryMethods/ucm113126.htm (accessed on 17 May 2018).
27. FDA's Center for Food Safety and Applied Nutrition. Laboratory Methods, Preparation and LC/MS/MS Analysis of Honey for Fluoroquinolone Residues. Available online: https://www.fda.gov/Food/FoodScienceResearch/LaboratoryMethods/ucm071495.htm (accessed on 17 May 2018).
28. U.S. Food and Drug Administration. Field Science and Laboratories, Laboratory Information Bulletins. Available online: https://www.fda.gov/ScienceResearch/FieldScience/ucm231463.htm (accessed on 17 May 2018).
29. FDA's Center for Food Safety and Applied Nutrition. Laboratory Methods, Bacteriological Analytical Manual (BAM). Available online: https://www.fda.gov/Food/FoodScienceResearch/LaboratoryMethods/ucm2006949.htm (accessed on 17 May 2018).
30. Horwitz, W. *Official Methods of Analysis of AOAC International*, 17th ed.; AOAC International: Gaithersburg, MD, USA, 2000.
31. Freitas, A.; Barbosa, J.; Ramos, F. Multi-residue and multi-class method for the determination of antibiotics in bovine muscle by ultra-high-performance liquid chromatography tandem mass spectrometry. *Meat Sci.* **2014**, *98*, 58–64. [CrossRef] [PubMed]
32. Freitas, A.; Leston, S.; Rosa, J.J.J.; Castilho, M.; Barbosa, J.; Rema, P.; Pardal, M.Â.A.; Ramos, F. Multi-residue and multi-class determination of antibiotics in gilthead sea bream (*Sparus aurata*) by ultra high-performance liquid chromatography-tandem mass spectrometry. *Food Addit. Contam. Part A* **2014**, *31*, 817–826. [CrossRef] [PubMed]
33. Charitonos, S.; Samanidou, V.F.; Papadoyannis, I. Development of an HPLC-DAD Method for the Determination of Five Sulfonamides in Shrimps and Validation According to the European Decision 657/2002/EC. *Food Anal. Methods* **2017**, *10*, 2011–2017. [CrossRef]

34. Konak, Ü.; Certel, M.; Şık, B.; Tongur, T. Development of an analysis method for determination of sulfonamides and their five acetylated metabolites in baby foods by ultra-high performance liquid chromatography coupled to high-resolution mass spectrometry (Orbitrap-MS). *J. Chromatogr. B Anal. Technol. Biomed. Life Sci.* **2017**, *1057*, 81–91. [CrossRef] [PubMed]
35. Liu, S.; Du, J.; Chen, J.; Zhao, H. Determination of 19 antibiotic and 2 sulfonamide metabolite residues in wild fish muscle in mariculture areas of Laizhou Bay using accelerated solvent extraction and high performance liquid chromatography-tandem mass spectrometry. *Se Pu Chin. J. Chromatogr.* **2014**, *32*, 1320–1325. [CrossRef]
36. Hoff, R.B.; Pizzolato, T.M.; Peralba, M.D.C.R.; Díaz-Cruz, M.S.; Barceló, D. Determination of sulfonamide antibiotics and metabolites in liver, muscle and kidney samples by pressurized liquid extraction or ultrasound-assisted extraction followed by liquid chromatography-quadrupole linear ion trap-tandem mass spectrometry (HPLC-QqL). *Talanta* **2015**, *134*, 768–778. [CrossRef] [PubMed]
37. Abdallah, H.; Arnaudguilhem, C.; Jaber, F.; Lobinski, R. Multiresidue analysis of 22 sulfonamides and their metabolites in animal tissues using quick, easy, cheap, effective, rugged, and safe extraction and high resolution mass spectrometry (hybrid linear ion trap-Orbitrap). *J. Chromatogr. A* **2014**, *1355*, 61–72. [CrossRef] [PubMed]
38. Huertas-Pérez, J.F.; Arroyo-Manzanares, N.; Havlíková, L.; Gámiz-Gracia, L.; Solich, P.; García-Campaña, A.M. Method optimization and validation for the determination of eight sulfonamides in chicken muscle and eggs by modified QuEChERS and liquid chromatography with fluorescence detection. *J. Pharm. Biomed. Anal.* **2016**, *124*, 261–266. [CrossRef] [PubMed]
39. Jia, W.; Shi, L.; Chu, X. Untargeted screening of sulfonamides and their metabolites in salmon using liquid chromatography coupled to quadrupole Orbitrap mass spectrometry. *Food Chem.* **2018**, *239*, 427–433. [CrossRef] [PubMed]
40. Li, J.; Liu, H.; Zhang, J.; Liu, Y.; Wu, L. A novelty strategy for the fast analysis of sulfonamide antibiotics in fish tissue using magnetic separation with high-performance liquid chromatography–tandem mass spectrometry. *Biomed. Chromatogr.* **2016**, *30*, 1331–1337. [CrossRef] [PubMed]
41. Huang, D.; Huang, X.; Gu, R.; Hui, Y.; Tian, L.; Feng, B.; Zhang, X.; Yu, H. Determination of 14 sulfonamide residues in shrimps by high performance liquid chromatography coupled with post-column derivatization. *Se Pu Chin. J. Chromatogr.* **2014**, *32*, 874–879. [CrossRef]
42. Freitas, A.; Barbosa, J.; Ramos, F. Multidetection of antibiotics in liver tissue by ultra-high-pressure-liquid-chromatography-tandem mass spectrometry. *J. Chromatogr. B Anal. Technol. Biomed. Life Sci.* **2015**, *976*, 49–54. [CrossRef] [PubMed]
43. Zhao, H.; Liu, H.; Yan, Z. Analysis of sulfonamide residues in pork and chicken by high performance liquid chromatography coupled with solid-phase extraction using multiwalled carbon nanotubes as adsorbent. *Se Pu Chin. J. Chromatogr.* **2014**, *32*, 294–298. [CrossRef]
44. Xia, L.; Liu, L.; Lv, X.; Qu, F.; Li, G.; You, J. Towards the determination of sulfonamides in meat samples: A magnetic and mesoporous metal-organic framework as an efficient sorbent for magnetic solid phase extraction combined with high-performance liquid chromatography. *J. Chromatogr. A* **2017**, *1500*, 24–31. [CrossRef] [PubMed]
45. Luo, H.; Xie, M.; Huang, X.; Wu, H.; Zhu, Z.; Huang, F.; Lin, X.; Ouyang, G. Multiresidue analysis of 63 veterinary drugs in meat by dispersive solid-phase extraction and high performance liquid chromatography-tandem mass spectrometry. *Chin. J. Chromatogr.* **2015**, *33*, 354–362. [CrossRef]
46. Wang, Z.; He, M.; Jiang, C.; Zhang, F.; Du, S.; Feng, W.; Zhang, H. Matrix solid-phase dispersion coupled with homogeneous ionic liquid microextraction for the determination of sulfonamides in animal tissues using high-performance liquid chromatography. *J. Sep. Sci.* **2015**, *38*, 4127–4135. [CrossRef] [PubMed]
47. Zhang, Q.; Xiao, X.; Li, G. Porous molecularly imprinted monolithic capillary column for on-line extraction coupled to high-performance liquid chromatography for trace analysis of antimicrobials in food samples. *Talanta* **2014**, *123*, 63–70. [CrossRef] [PubMed]
48. Huang, X.; Chen, L.; Yuan, D. Development of monolith-based stir bar sorptive extraction and liquid chromatography tandem mass spectrometry method for sensitive determination of ten sulfonamides in pork and chicken samples. *Anal. Bioanal. Chem.* **2013**, *405*, 6885–6889. [CrossRef] [PubMed]
49. Nebot, C.; Regal, P.; Miranda, J.M.; Fente, C.; Cepeda, A. Rapid method for quantification of nine sulfonamides in bovine milk using HPLC/MS/MS and without using SPE. *Food Chem.* **2013**, *141*, 2294–2299. [CrossRef] [PubMed]

50. Arroyo-Manzanares, N.; Gámiz-Gracia, L.; García-Campaña, A.M. Alternative sample treatments for the determination of sulfonamides in milk by HPLC with fluorescence detection. *Food Chem.* **2014**, *143*, 459–464. [CrossRef] [PubMed]
51. Wang, J.; Leung, D.; Chow, W.; Chang, J.; Wong, J.W. Development and Validation of a Multiclass Method for Analysis of Veterinary Drug Residues in Milk Using Ultrahigh Performance Liquid Chromatography Electrospray Ionization Quadrupole Orbitrap Mass Spectrometry. *J. Agric. Food Chem.* **2015**, *63*, 9175–9187. [CrossRef] [PubMed]
52. Sereshti, H.; Khosraviani, M.; Sadegh Amini-Fazl, M. Miniaturized salting-out liquid-liquid extraction in a coupled-syringe system combined with HPLC-UV for extraction and determination of sulfanilamide. *Talanta* **2014**, *121*, 199–204. [CrossRef] [PubMed]
53. Lu, Y.; Cong, B.; Tan, Z.; Yan, Y. Synchronized separation, concentration and determination of trace sulfadiazine and sulfamethazine in food and environment by using polyoxyethylene lauryl ether-salt aqueous two-phase system coupled to high-performance liquid chromatography. *Ecotoxicol. Environ. Saf.* **2016**, *133*, 105–113. [CrossRef] [PubMed]
54. Shao, M.; Zhang, X.; Li, N.; Shi, J.; Zhang, H.; Wang, Z.; Zhang, H.; Yu, A.; Yu, Y. Ionic liquid-based aqueous two-phase system extraction of sulfonamides in milk. *J. Chromatogr. B Anal. Technol. Biomed. Life Sci.* **2014**, *961*, 5–12. [CrossRef] [PubMed]
55. Hou, X.L.; Chen, G.; Zhu, L.; Yang, T.; Zhao, J.; Wang, L.; Wu, Y.L. Development and validation of an ultra high performance liquid chromatography tandem mass spectrometry method for simultaneous determination of sulfonamides, quinolones and benzimidazoles in bovine milk. *J. Chromatogr. B Anal. Technol. Biomed. Life Sci.* **2014**, *962*, 20–29. [CrossRef] [PubMed]
56. Kechagia, M.; Samanidou, V.; Kabir, A.; Furton, K.G. One-pot synthesis of a multi-template molecularly imprinted polymer for the extraction of six sulfonamide residues from milk before high-performance liquid chromatography with diode array detection. *J. Sep. Sci.* **2017**, *41*, 723–731. [CrossRef] [PubMed]
57. Ibarra, I.S.; Miranda, J.M.; Rodriguez, J.A.; Nebot, C.; Cepeda, A. Magnetic solid phase extraction followed by high-performance liquid chromatography for the determination of sulphonamides in milk samples. *Food Chem.* **2014**, *157*, 511–517. [CrossRef] [PubMed]
58. Tolmacheva, V.V.; Apyari, V.V.; Furletov, A.A.; Dmitrienko, S.G.; Zolotov, Y.A. Facile synthesis of magnetic hypercrosslinked polystyrene and its application in the magnetic solid-phase extraction of sulfonamides from water and milk samples before their HPLC determination. *Talanta* **2016**, *152*, 203–210. [CrossRef] [PubMed]
59. Li, Y.; Wu, X.; Li, Z.; Zhong, S.; Wang, W.; Wang, A.; Chen, J. Fabrication of $CoFe_2O_4$-graphene nanocomposite and its application in the magnetic solid phase extraction of sulfonamides from milk samples. *Talanta* **2015**, *144*, 1279–1286. [CrossRef] [PubMed]
60. Karageorgou, E.; Manousi, N.; Samanidou, V.; Kabir, A.; Furton, K.G. Fabric phase sorptive extraction for the fast isolation of sulfonamides residues from raw milk followed by high performance liquid chromatography with ultraviolet detection. *Food Chem.* **2016**, *196*, 428–436. [CrossRef] [PubMed]
61. Chatzimitakos, T.; Samanidou, V.; Stalikas, C.D. Graphene-functionalized melamine sponges for microextraction of sulfonamides from food and environmental samples. *J. Chromatogr. A* **2017**, *1522*, 1–8. [CrossRef] [PubMed]
62. Gao, M.; Yan, H.; Sun, N. Water-compatible poly (hydroxyethyl methacrylate) polymer sorbent for miniaturized syringe assisted extraction of sulfonamides in milk. *Anal. Chim. Acta* **2013**, *800*, 43–49. [CrossRef] [PubMed]
63. Zhao, H.; Zulkoski, J.; Mastovska, K. Development and Validation of a Multiclass, Multiresidue Method for Veterinary Drug Analysis in Infant Formula and Related Ingredients Using UHPLC-MS/MS. *J. Agric. Food Chem.* **2017**, *65*, 7268–7287. [CrossRef] [PubMed]
64. Pérez, M.L.G.; Romero-González, R.; Vidal, J.L.M.; Frenich, A.G. Analysis of veterinary drug residues in cheese by ultra-high-performance LC coupled to triple quadrupole MS/MS. *J. Sep. Sci.* **2013**, *36*, 1223–1230. [CrossRef] [PubMed]
65. Wu, L.; Song, Y.; Hu, M.; Xu, X.; Zhang, H.; Yu, A.; Ma, Q.; Wang, Z. Determination of sulfonamides in butter samples by ionic liquid magnetic bar liquid-phase microextraction high-performance liquid chromatography. *Anal. Bioanal. Chem.* **2015**, *407*, 569–580. [CrossRef] [PubMed]

66. Summa, S.; Lo Magro, S.; Armentano, A.; Muscarella, M. Development and validation of an HPLC/DAD method for the determination of 13 sulphonamides in eggs. *Food Chem.* **2015**, *187*, 477–484. [CrossRef] [PubMed]
67. He, X.; Tan, L.; Wu, W.; Wang, J. Determination of sulfadiazine in eggs using molecularly imprinted solid-phase extraction coupled with high-performance liquid chromatography. *J. Sep. Sci.* **2016**, *39*, 2204–2212. [CrossRef] [PubMed]
68. Sajid, M.; Na, N.; Safdar, M.; Lu, X.; Ma, L.; He, L.; Ouyang, J. Rapid trace level determination of sulfonamide residues in honey with online extraction using short C-18 column by high-performance liquid chromatography with fluorescence detection. *J. Chromatogr. A* **2013**, *1314*, 173–179. [CrossRef] [PubMed]

Article

Development and Validation of an HPLC-DAD Method for the Simultaneous Extraction and Quantification of Bisphenol-A, 4-Hydroxybenzoic Acid, 4-Hydroxyacetophenone and Hydroquinone in Bacterial Cultures of *Lactococcus lactis*

Angelos T. Rigopoulos [1], Victoria F. Samanidou [2] and Maria Touraki [1,*]

1 Laboratory of General Biology, Division of Genetics, Development and Molecular Biology, Department of Biology, School of Sciences, Aristotle University of Thessaloniki (A.U.TH.), 54 124 Thessaloniki, Greece; angelos.rigopoulos@embl.de
2 Laboratory of Analytical Chemistry, Department of Chemistry, School of Sciences, Aristotle University of Thessaloniki (A.U.TH.), 54 124 Thessaloniki, Greece; samanidu@chem.auth.gr
* Correspondence: touraki@bio.auth.gr; Tel.: +30-231-099-8292

Received: 8 January 2018; Accepted: 31 January 2018; Published: 6 February 2018

Abstract: Bisphenol-A, a synthetic organic compound with estrogen mimicking properties, may enter bloodstream through either dermal contact or ingestion. Probiotic bacterial uptake of bisphenol can play a major protective role against its adverse health effects. In this paper, a method for the quantification of BPA in bacterial cells of *L. lactis* and of BPA and its potential metabolites 4-hydroxybenzoic Acid, 4-hydroxyacetophenone and hydroquinone in the culture medium is described. Extraction of BPA from the cells was performed using methanol–H_2O/TFA (0.08%) (5:1 v/v) followed by SPE. Culture medium was centrifuged and filtered through a 0.45 μm syringe filter. Analysis was conducted in a Nucleosil column, using a gradient of A (95:5 v/v H_2O: ACN) and B (5:95 v/v H_2O: ACN, containing TFA, pH 2), with a flow rate of 0.5 mL/min. Calibration curves (0.5–600 μg/mL) were constructed using 4-*n*-Octylphenol as internal standard ($1 > R^2 > 0.994$). Limit of Detection (LOD) and Limit of Quantification (LOQ) values ranged between 0.23 to 4.99 μg/mL and 0.69 to 15.1 μg/mL respectively. A 24 h administration experiment revealed a decline in BPA concentration in the culture media up to 90.27% while the BPA photodegradation levels were low. Our results demonstrate that uptake and possible metabolism of BPA in *L. lactis* cells facilitates its removal.

Keywords: HPLC-DAD; bisphenol A; 4-Hydroxybenzoic Acid, 4-Hydroxyacetophenone hydroquinone; *Lactococcus lactis*

1. Introduction

Bisphenol-A is the organic synthetic compound 4, 4′-Isopropylidenediphenol, which is widely used in thermal paper industry and as a component of synthetic plastic (vinyl-chloride) due to its mechanical and extreme-temperature resistance [1]. BPA, was found to exhibit estrogenic properties, arising from its structural resemblance to the human 17β-estradiol [2], thus it is also referred to as xenoestrogen or endocrine disruptor. The main routes of human exposure to BPA are the consumption of food in BPA containing packages [3] and through dermal contact with BPA rich materials [4]. Following exposure, BPA is mainly metabolized in humans to BPA-glucuronide through the hepatic glucuronide transferase and excreted from the body [5,6]. However, a percentage of BPA, referred to as "free" or "active" BPA, remains in the blood circulation for up to a week and can interfere with endogenous biological processes [4]. Due to its estrogenic and genotoxic effects and its wide occurrence, BPA presents a risk to humans and animals and hence it is the object of great environmental concern [7–9].

Although part of bisphenol is degraded via photo-oxidation [10–12], monitoring of its levels is deemed necessary in various environmental sources [13]. To this end, biodegradation has been proposed as an advanced technique for BPA elimination [13–15], since a variety of microorganisms have been identified to decompose BPA [15,16]. Probiotics are bacteria beneficial to the host, which have been recognized as safe for human consumption and are used in the production of food products [17,18]. Recently they have been reported to be ideal against heavy metal toxicity since they bind and sequester metals [19]. Moreover, the probiotic bacteria *Bifidobacterium breve* and *Lactobacillus casei* when administered to rats showed a prophylactic effect against the detrimental effects of BPA, by reducing its intestinal absorption and facilitating its excretion, [18]. Likewise, the ability of *Bacillus* strains [20–22], certain *Lactococcus* strains [23] and *Shingomonas paucimobilis* [24] to degrade bisphenol has been documented. Recently a simplified method for the extraction of bisphenol from bacterial culture suspensions has been reported [25]. Moreover, the compounds 4-hydroxy acetophenone (HAP), 4-hydroxy benzoic acid (HBA) and Hydroquinone (HQ) have been proposed as major bacterial metabolites of BPA bacterial degradation pathways [14,26,27]. However, all these studies employed the determination of bisphenol in the cell free culture supernatant and not in the bacterial cells even though they examine the kinetics of bisphenol.

Analytical methods for the determination of bisphenol in various food samples [3], as well as in biological fluids and environmental samples [28], have been extensively reviewed. The determination of BPA has also been investigated in more complex matrices such as rat tissues [29], human breast milk [30] resin-based dental restorative materials [31] and saliva [32]. However, to the best of our knowledge, no method has been reported for the determination of bisphenol in bacterial cells.

In the present study, the development and validation of an analytical method for the quantification of BPA in bacterial cells is reported for the first time using the widespread probiotic bacterium *Lactococcus lactis* (ATC 11454). The method is also validated in bacterial culture supernatants for the simultaneous determination of the bacterial metabolites reported to be produced during the bacterial major metabolic pathway [14], namely 4-hydroxy-acetophenone (HAP), 4-hydroxy benzoic acid (HBA) and Hydroquinone (HQ). The application of this method demonstrates the uptake of bisphenol by *Lactococcus lactis* cells and allows its quantification.

2. Materials and Methods

2.1. Chemicals and Reagents

Acetonitrile was purchased from VWR Chemicals (Paris, France). Methanol and Absolute ethanol were purchased from Fisher Scientific (Loughborough, UK). Trifluoroacetic acid (TFA), KH_2PO_4 and $FeCl_3$ were purchased from AppliChem GmbH (Darmstadt, Germany), while NaCl, $CaCl_2$, $(NH_4)_2SO_4$ and $MgSO_4$ were purchased from Merck KGaA (Darmstadt, Germany). All the chemicals used were HPLC Grade, unless specified otherwise.

Bisphenol-A (BPA) and the internal standard 4-*n*-Octylphenol (OP) were purchased from Alfa Aesar (Karlsruhe, Germany). 4-Hydroxyacetophenone (HAP), 4-Hydroxybenzoic Acid (HBA) and Hydroquinone (HQ) were purchased from Acros Organics (Ceel, Belgium).

The bacterial strain from which the liquid culture supernatant and the pellet originated and used for the spike-recovery studies, was *Lactococcus lactis* (ATCC 11454). The liquid culture medium that was used for the growth of *L. lactis* was Brain Heart Infusion broth (BHI broth) and was purchased from AppliChem GmbH (Darmstadt, Germany). An artificial bacterial culture medium was prepared, namely Basal Salts Medium (BSM), composed of KH_2PO_4 and $(NH_4)_2SO_4$ 1 g/L each, $MgSO_4$ 0.1 g/L, NaCl and $CaCl_2$ 0.05 g/L each and $FeCl_3$ 0.01 g/L. BPA was used as a sole carbon source in BSM in varying concentrations for the feeding experiments. For the retrieval of the bacterial culture supernatant and pellet 100 μL of an *L. lactis* glycerol stock solution were inoculated in a 250 mL flash with BHI broth. Following a 24-h incubation at 28 °C under constant stirring, the culture was centrifuged at 5000 rpm and the precipitate was transferred to a 250 mL flask containing BSM. At the end of the experiment, the culture was centrifuged, the precipitate was placed on a filtering paper and its weight was recorded following removal of excess water. The recorded weight is referred as 'wet cell weight'. The culture supernatant and the wet cell weight were used for the spike-recovery assays.

Solid Phase Extraction columns, frits and C_{18} sorbent material, 35–75 U were purchased from Grace Davison Discovery Sciences (Bannockburn, IL, USA). Membrane filters with 0.45 μm diameter pores were purchased from Merck KGaA (Darmstadt, Germany).

2.2. Standard Preparation

2.2.1. Stock Solution Preparation

Stock solutions 500 μg/mL of BPA, HAP, HBA and HQ were prepared by dissolving 25 mg of the analyte in 50 mL methanol. The stock solution of BPA (50 mg/mL) was prepared by dissolving 2.5 g of BPA in 50 mL ethanol. The internal standard stock solution of 500 μg/mL was prepared by dissolving 25 mg of OP in 50 mL methanol.

2.2.2. Working Standard Solution Preparation

Seven-point and six-point working standard solutions were prepared for analytes at concentrations of 0.5, 10, 50, 100, 200, 400, and 600 μg/mL for BPA, at 0.5, 2, 4, 6, 8 and 10 μg/mL for HAP and HQ and at 1, 2, 4, 6, 8 and 10 μg/mL for HBA and the internal standard 4-*n*-octylphenol (75 μg/mL).

2.3. Sample Preparation

For the bacterial pellet method development, 0.1 g of wet cell weight *L. lactis* was added in a test tube with 8 mL methanol, 2 mL of double distilled water containing 0.08% (*v*/*v*) TFA and 150 μL internal standard from the 500 μg/mL stock. The concentration of the internal standard in the final 1 mL volume was 75 μg/mL. The test tube was transferred to an ice-filled container and was placed under a homogenizer for 5 min. The homogenate was separated in ten Eppendorf centrifuge tubes and was centrifuged for 10 min in 8000 rpm. The supernatant was collected and mixed with an equal volume of ddH_2O containing 0.08% TFA. BPA extraction was performed using Solid Phase Extraction (SPE). An SPE cartridge was prepared using 0.5 g of C_{18} sorbent in an SPE column. The SPE cartridge was conditioned with 10 mL methanol, 5 mL ddH_2O and 5 mL ddH_2O containing 0.08% TFA. The sample was loaded on the cartridge and a manual flow of 1 mL/min was employed. The cartridge was, then, washed with 5 mL ddH_2O and elution was performed using 4 mL methanol. The eluate was collected and placed under a rotary evaporator in a 37 °C water bath, to dryness. Finally the sample was reconstituted with 1 mL acetonitrile and stored in an aluminum foil-covered Eppendorf tube at −20 °C. All experiments were conducted in dark or covered tubes to avoid photodegradation of BPA.

For the bacterial supernatant, an 8.5 mL sample of the culture was added in a test tube along with 1.5 mL internal standard from the stock solution. The sample was placed in 10 Eppendorf centrifuge tubes (1 mL per tube), and was centrifuged for 10 min in 8000 rpm. The supernatant was collected and filtered through a 0.45 μm diameter pore filter. The filtrate was stored in an aluminum foil-covered Eppendorf tube at −20 °C.

2.4. BPA Administration (Feeding) Assay

For the feeding assays 6 falcon tubes of 50 mL volume were filled with BHI broth and were inoculated with 100 μL of a *L. lactis* glycerol stock solution. The cultures were incubated for 24 h at 28 °C under constant stirring. The cultures were, then, centrifuged and the medium was discarded. A total of 50 mL of fresh BSM was added to each culture along with varying quantities from the BPA stock solution (50 mg/mL) so that the final BPA concentration in each culture would be 50, 100, 150, 200, 400 και 500 μg/mL respectively. The cultures were incubated for another 24 h at 28 °C under constant stirring. The remaining BPA in the cultures was quantified using the method and the calibration curves that were described.

2.5. Photodegradation Assay

To investigate potential photodegradation of BPA, a falcon tube of 50 mL containing BSM and BPA at a concentration of 50 μg/mL was prepared and was incubated under the same conditions as the feeding assays. BPA was quantified at the beginning and the end of the 24 h incubation period. The presence of photodegradation products was examined in the resulting chromatograms.

2.6. HPLC-DAD Conditions and Methods

The instrument used consisted of an $LC20_{AD}$ pump and an SPD-20A photodiode array detector (DAD) purchased from Shimadzu (Kyoto, Japan). Separation of BPA was achieved on a Nucleosil 100 C_{18} column (250 × 4.6 mm, 5 μm) purchased by Macherey-Nagel GmbH & Co. (Duren, Germany) with a binary mobile phase system at room temperature. The injection was performed using a 100 μL Hamilton syringe, with an injection volume of 80 μL. Mobile Phase A consisted of a 0.1% TFA in ddH_2O solution and Acetonitrile at a ratio of 95:5, *v*/*v*. Mobile Phase B consisted of a 0.08% TFA in ddH_2O solution and Acetonitrile at a ratio of 5:95, *v*/*v*. The gradient started at 100% Mobile Phase A which was brought down to 80% over 1 min, 70% over 4 min, 55% at 15 min, 40% at 17 min, 20% at 20 min and finally to 10% over 2 min and then was kept at this level until the end on the analysis. The flow rate was constant at 0.5 mL/min. BPA was detected at 220 nm with a band width of 2.0 nm. HAP, HBA and HQ were detected at 280 nm with a band width of 2.0 nm. Identification was conducted by matching the peak retention times (*Rt*) with those of the pure standards. Internal standard calibration was used by plotting the peak area ratio of the analyte to the peak area of the internal standard against the concentration of the analyte.

3. Results and Discussion

3.1. Optimization of Sample Preparation

The homogenization of the sample was performed to lyse the cells and release the adsorbed BPA for the feeding assay. Initially, ethyl acetate was used instead of methanol and a manual homogenization was attempted with a glass homogenizer, but cell lysis was deemed inadequate. Hence methanol was selected as the extraction solvent. The addition of an equal volume ddH_2O containing 0.08% (*v*/*v*) TFA after centrifugation aimed to increase the solubility of BPA due to further dilution.

An elaborate and efficient vortex assisted liquid-liquid microextraction with octanol was used for the trace analysis of bisphenol in water and wastewater samples [33]. However, an attempt to use liquid-liquid extraction with ethyl acetate in the samples of the present study resulted to the formation of an emulsion zone between the two phases, a fact that led to greatly varying percentages of BPA recovery due to the loss of the analyte during the procedure. Thus, solid phase extraction was implemented, since SPE is a common step in the extraction procedures of bisphenol from complex matrices, such as rat tissues [29] or human serum [5] and its benefits in the biomonitoring of bisphenol have been demonstrated [28,33]. In the present study a flow rate of 1 mL/min was selected throughout the SPE procedure, as effective and timely favorable, while higher flow rates did not provide full retention of BPA by the C_{18} material, especially at high BPA concentrations employed and resulted in lower recovery values. A 37 °C water bath was selected to facilitate the evaporation of methanol after SPE. The slightly higher evaporation temperatures of 45 °C that have been reported for the analysis of bisphenol and 4-nonylphenol [29] resulted in heat-induced decomposition of the internal standard when used in the present study. Our results for BPA extraction are in accordance with a recently published report on a methanol extraction of bisphenols from bacterial cultures and solid matrices [25] in which, however, SPE was not employed and lower concentrations of BPA were used [25], resulting in recoveries up to 95%. The SPE step used in the present study greatly improved recovery which amounted to 99.38–100.18% in culture supernatants and to 90.8–102.46% in bacterial cells, in the wide range of concentrations employed (Table 1).

Table 1. Analyses of the calibration curves for analytes in the cell culture supernatant and the cell culture precipitate.

Analyte in Media/Std. Curve Equation/R^2	Nominal Conc. (μg/mL)	Mean Calculated Concentration (μg/mL) (Mean ± SD)	Relative Bias (%)	Precision		Recovery Rate (Mean ± SD)
				Intra-Day (RSD %)	Inter-Day (RSD %)	
BPA in BSM Supernatant y = 0.0295x + 0.0077 1	0.5	0.48 ± 0.05	−3.6	8.7	10.4	96 ± 13
	10	9.94 ± 0.08	−0.6	0.1	0.8	99.4 ± 0.7
	50	50.2 ± 0.7	0.4	0.8	1.4	100 ± 1
	100	100 ± 2	0.05	0.8	1.7	100 ± 1
	200	200 ± 2	−0.2	0.7	1.2	100 ± 1
	400	401 ± 3	0.2	0.2	0.8	100.2 ± 0.6
	600	600 ± 3	−0.01	0.3	0.6	100.0 ± 0.5
BPA in *L. lactis* precipitate y = 0.0232x + 0.0092 0.9999	0.5	0.41 ± 0.03	−18.5	3.2	7.4	82 ± 5
	10	9.1 ± 0.4	−9.2	2.3	4.7	91 ± 4
	50	48± 1	−3.7	2.7	2.7	96 ± 2
	100	98 ± 3	−2.0	2.0	2.9	98 ± 2
	200	205 ± 3	2.5	0.4	1.4	103 ± 1
	400	400.6 ± 0.9	0.2	0.2	0.2	100 ± 0.2
	600	597 ± 5	−0.6	0.1	0.8	99.4 ± 0.7
HAP in BSM y = 0.202x + 0.0134 0.9994	0.5	0.52 ± 0.02	3.4	2.8	3.8	103 ± 3
	2	2.06 ± 0.05	2.9	0.2	2.4	103 ± 2
	4	3.84 ± 0.04	−4.1	0.7	1.0	95.9 ± 0.8
	6	6.1 ± 0.1	1.1	1.0	2.3	101 ± 2
	8	8.0 ± 0.2	0.6	2.1	2.2	101 ± 2
	10	10.0 ± 0.1	−0.3	0.3	1.0	99.7 ± 0.8
HBA in BSM y = 0.0407x + 0.0073 0.994	1	0.91 ± 0.03	−9.5	2.7	3.3	90 ± 2
	2	1.86 ± 0.04	−6.9	0.3	2.2	93 ± 3
	4	4.1 ± 0.2	2.3	3.2	4.2	102 ± 2
	6	6.2 ± 0.1	3.0	2.1	2.3	103 ± 2
	8	8.3 ± 0.3	4..2	3.0	3.1	104 ± 3
	10	9.62 ± 0.04	−3.8	0.3	0.4	96.2 ± 0.3
HQ in BSM y = 0.0351x − 0.0082 0.999	0.5	0.48 ± 0.01	−3.8	1.4	2.1	96 ± 2
	2	2.2 ± 0.1	7.5	3.4	6.5	108 ± 5
	4	3.91 ± 0.03	−2.2	0.03	0.8	97.8 ± 0.6
	6	5.91 ± 0.09	−1.5	1.6	1.5	98 ± 1
	8	7.9 ± 0.2	−1.4	0.5	2.0	99 ± 2
	10	10.1 ± 0.2	1.3	1.4	2.2	101 ± 2

In an effort to perform the analysis of BPA in the BHI medium, which is recommended for growth of *L. lactis*, the nutrients in the broth resulted in the presence of peaks in the chromatogram that appeared to interfere with the tested bacterial metabolites peaks. The artificial bacterial medium (BSM) was prepared to tackle the problems that arose in the analysis using the reference BHI medium. Moreover, in this manner BPA was the sole carbon source for the bacterium, a fact that contributes in its biotransformation.

3.2. Method Validation

The results of the linear regression analysis performed using the "Linear Regression" package of Microsoft Excel 2013 software are presented in Table 1, while detection limits and Retention time variance are presented in Table 2. The coefficient of determination values (R^2) for the standard curves in BSM supernatant and *L. lactis* cells were 1 and 0.999 respectively, while for HAP, HBA and HQ in BSM supernatant the values exceeded 0.9994, 0.994 and 0.999 respectively, indicating excellent linearity. The Relative Bias was calculated by computing the percentage of the difference of the mean calculated concentration to the nominal concentration divided by the nominal concentration of the analyte. All results are within the accepted range of ±10%. Intra-day precision was calculated by performing a total of three injections of the nominal concentration on the same day, while inter-day precision was calculated by performing six injections of the nominal concentration over three days. Both were computed as % R.S.D. = (S.D./mean) × 100. None of the values exceeded 6.0% for all analytes, with the exception of the values obtained for the lowest nominal value of 0.5 μg/mL of BPA in bacterial cells and in the medium, which might be due to the fact that this nominal value is below LOD (Table 2). Since an elaborate extraction protocol was required for the bacterial cells, the use of an internal standard was deemed necessary, to compensate for sample losses during sample preparation. The internal standard 4-*n*-octylphenol was selected, due to the fact that it is well resolved and it has a structure similar to bisphenol. Moreover, it is not expected to be present in samples, since only 4-*t*-octyphenol has been reported to be found at very low concentrations in environmental samples and human milk [34,35]. Our results, as presented in Table 1, indicate that the method is highly precise and repeatable. Recovery rates were well within range of 10% error, reaching close to 100% recovery, thus further supporting the precision of this method. The recovery results presented in Table 1 are in good correlation with those reported for rat serum (95%) and higher than those reported for rat testis (78.6%) and liver (84.0%) [29].

Table 2. Analyses LOD, LOQ and *Rt* values for all analytes.

Std. Curve	Retention Time (Min) Mean ± SD (RSD %)	LOD (μg/mL)	LOQ (μg/mL)
BPA in *L. lactis*	26.68 ± 0.13 (0.41)	4.99	15.01
BPA in BSM	26.71 ± 0.09 (0.38)	0.64	1.93
HAP in BSM	15.66 ± 0.05 (0.29)	0.24	0.73
HBA in BSM	12.77 ± 0.12 (0.23)	0.73	2.21
HQ in BSM	10.80 ± 0.08 (0.90)	0.32	0.98

The nominal concentrations of 0.5 μg/mL BPA in BSM and cell precipitate were below the calculated LOD values (Table 2) and were thus excluded from the estimation of recovery results described above. The values for the Limit of Detection (LOD) and Limit of Quantification (LOQ) were calculated from the calibration curves. The LOD was calculated to 3.3 times the ratio of the Standard Error (intercept) to the Coefficients x, while the LOQ was 10 times this ratio. The retention time means, standard deviations and RSDs were calculated using the data from one chromatogram of each of the nominal concentrations used. The value of an LOD of 4.99 μg/mL estimated in the present study for BPA in bacterial cells at the range of 10–600 μg/mL, is higher than the value of 2.8 ng/mL reported for rat tissues, but over a narrower range of concentrations of 0.15–150 μg BPA/mL [29]. A lower value of LOD was found for BPA in the less complex sample of BSM, amounting to 0.64 μg/mL (Table 2), although the same range of concentrations was employed (Table 1), possibly due to lower matrix complexity. Similar low LOD values were estimated for all possible metabolites tested, as shown in Table 2. Although these compounds have been proposed as possible bisphenol bacterial metabolites, they were identified using MS and NMR analysis of the HPLC isolated fractions [27]. In fact, in the studies on the determination of bisphenol degradation using HPLC-DAD [20,23], no validation was included. The method presented in this work allows the simultaneous quantification of bisphenol A and three of its major bacterial metabolites.

3.3. Feeding Results

A typical chromatogram of HQ, HBA and HAP and BPA is presented in Figure 1. Analyte peaks of HQ, HBA, HAP, BPA and internal standard OP are shown with a retention time (*Rt*) of 10.671, 12.728, 15.579, 26.241 and 40.771 min respectively. The peak that appears at 25.6 (Figure 1) might be attributed to a BPA photodegradation product, since it was also identified during the photodegradation experiment and is presented in Figure 2. Photodegradation of BPA via photooxidation has been confirmed [10,12], and the products of this process presented higher toxicity than the parent compound [11]. Since the feeding experiments were conducted under normal light, it was deemed appropriate to test if any BPA loss was due to photodegradation. In these experiments the initial BPA concentration was calculated at 48.65 μg/mL and the concentration following the 24 h incubation amounted to 44.18 μg/mL. Hence, the BPA loss due to photodegradation was 9.19%. These values are in accordance with those reported for bisphenol A degradation by UV irradiation in the absence of hydrogen peroxide [36].

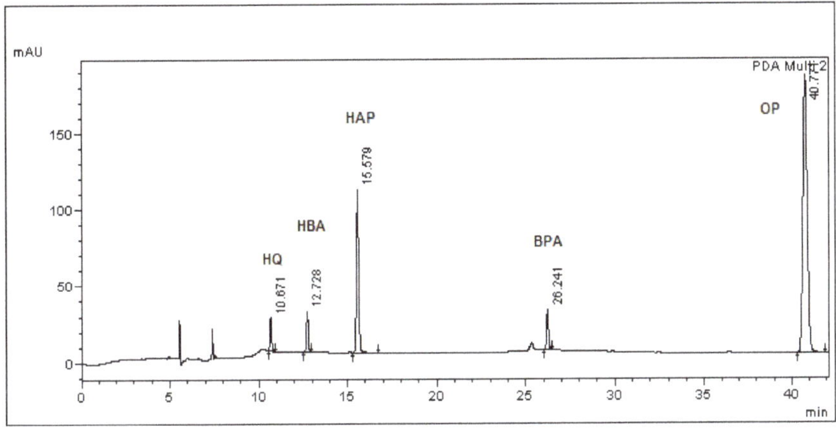

Figure 1. A chromatogram containing all analytes at a concentration of 10 μg/mL and internal standard at 75 μg/mL in fresh BSM without *L. lactis* inoculation. The first two peaks with *Rt* of 5.7 min and 7.5 min are background noise from acetonitrile solvent and TFA.

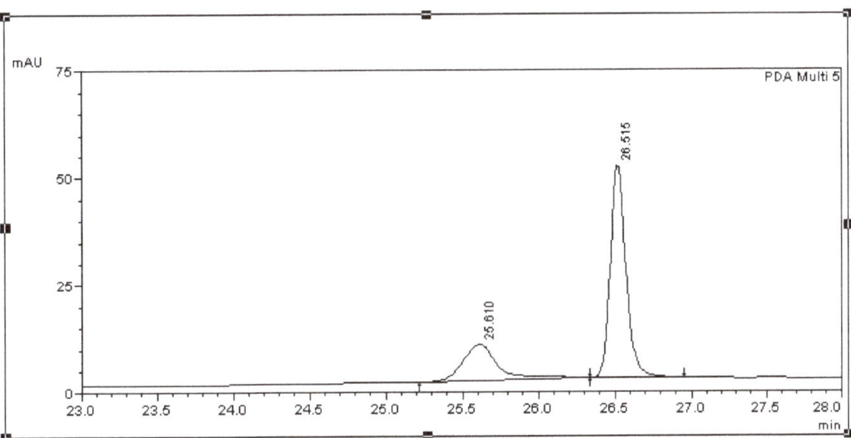

Figure 2. Chromatogram displaying BPA and its identified photodegradation product with a *Rt* of 26.515 min and 25.610 min respectively. The compounds were detected at a wavelength of 300 nm.

Typical chromatograms of the feeding assays are presented in Figure 3 (bacterial cells) and Figure 4 (bacterial culture medium). The relative blank chromatograms resulting from bacterial cells and from their culture medium any additions are shown in the Supplemental section. The peaks appearing at 10.2 min in both chromatograms does not represent one of the metabolites described in this paper, as confirmed by comparing the absorbance spectra of the peak to those of the pure metabolites analyzed.

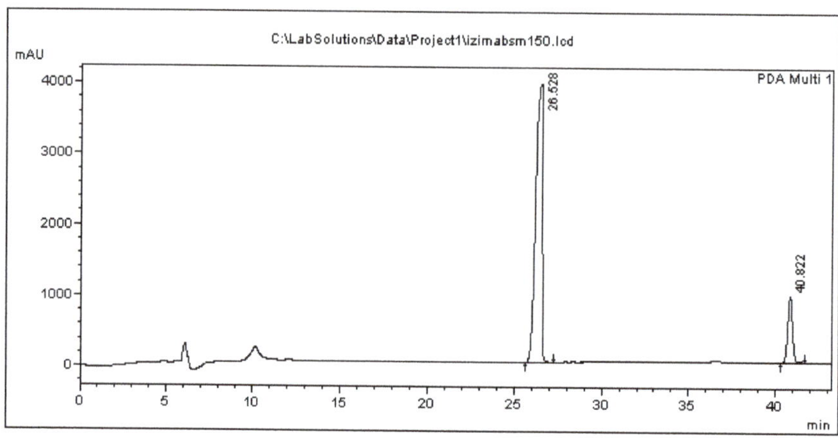

Figure 3. Chromatogram of cell precipitate originating from the administration of 150 μg/mL BPA in the bacterial cells. The peaks at 26.528 min and 40.822 min correspond to BPA and the internal standard.

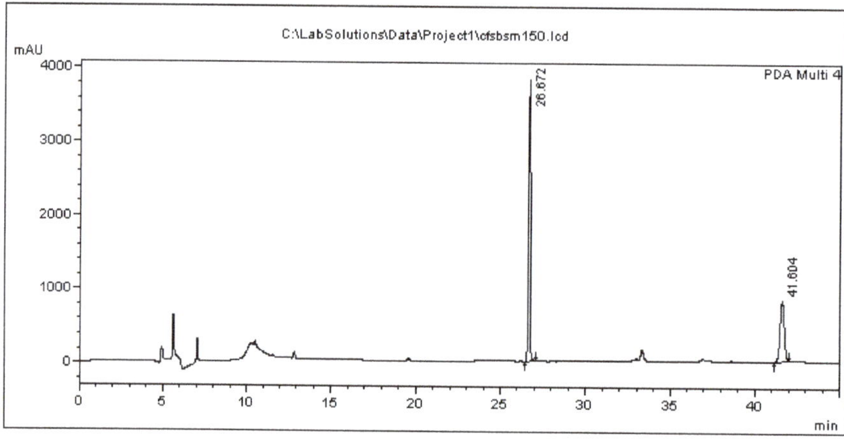

Figure 4. Chromatogram of culture supernatant originating from the administration of 150 μg/mL BPA in the bacterial cells. The peaks at 26.672 min and 41.604 min correspond to BPA and the internal standard.

Two major pathways have been proposed for the metabolism of bisphenol A by bacteria, the first leading to mineralization, with HBA and HAP as intermediates and the second to hydroxylation [6,16]. However, it has been pointed out that the metabolic pathways might be strain specific [14], since certain metabolites produced by one strain are not detected when another bacterial strain is used. In a similar manner, the analytes HQ, HBA and HAP were not detected in the chromatograms of the feeding assays. Absence of bacterial metabolites has also been reported for *Bacillus pumilus* [20] and certain *Lactococcus* strains [23]. In all these reports, including the present work, the presence of metabolites was examined at 24 h of culture. However, when various *Bacilli* and *Sphingomonas* strains were tested for bisphenol degrading capacity, metabolites were detected for three of them and only from 8–12 h, in a declining manner [37]. These results indicate that the time period of 24 h used in the present paper might be too long, since the metabolites under investigation might have formed and then further transformed. On the other hand, the bacterial metabolites examined in the present paper were reported to be produced by *Bacillus* sp., yet only after 3 days of incubation in the presence of 20 mM BPA [21]. Moreover, the photodegradation of BPA metabolites has been reported for hydroxybenzoic acid as well as for hydroquinone [38,39]. It is evident that more experiments are needed in the course of bacterial incubation, since metabolism is an ongoing process and each strain presents particular metabolic characteristics.

A depletion of up to 58% in total BPA (μg) was observed in our experiments, in all bacterial cultures as shown in Table 3. Complete degradation of BPA has been reported following a 48 h incubation period [37] or after 9 days of incubation [21]. In our experiments BPA was reduced by 90.27%, 64.16%, 58.60%, 67.74%, 79.83% and 85.05% in the cultures corresponding to initial BPA concentration of 50, 100, 150, 200, 400 and 500 μg/mL. Our results suggest that the initial BPA concentration added has a profound effect on its degradation. Many bacterial strains that can degrade BPA have been reported, and this degrading ability varies among the species [14]. It should also be noted that different initial concentrations of BPA are used in the reports employing bacteria, ranging from 0.1 to 60 mg/L [22,37,40] and for varying incubation periods. Bacteria can clearly employ a variety of metabolic pathways to exploit pollutants as energy sources, under the prerequisite that the pollutants are bioavailable and at levels that could trigger such mechanisms [16].

Table 3. Comparative data of BPA administered (μg) and found in feeding assays.

BPA Administered in 50 mL Culture (μg)	BPA Found in Precipitate	BPA Found in 50 mL Supernatant	Total BPA Found	Concentration Decline in Cultures (%)
		(μg)		
2500 (50 μg/mL)	46.77	196.46	243.23	90.27
5000 (100 μg/mL)	110.48	1681.74	1792.22	64.16
7500 (150 μg/mL)	255.75	2849.48	3105.23	58.60
10,000 (200 μg/mL)	262.20	2964.01	3226.22	67.74
20,000 (400 μg/mL)	437.24	3595.83	4033.07	79.83
25,000 (500 μg/mL)	421.16	3316.41	3737.57	85.05

4. Conclusions

The developed analytical method provides a reliable, robust and sensitive approach for the simultaneous detection and quantification of bisphenol A and its potential metabolites in the complex matrix sample of bacterial cells. Good chromatography separation was achieved on a Nucleosil 100 C_{18} column with a longer run time, which however allows the detection and separation of metabolites. LC-DAD combined with solid phase extraction and the use of 4-*n*-octylphenol as the internal standard, is capable of operating with a detection limit of 4.99 μg/mL for bisphenol and lower limits for metabolites in bacterial cells. The method was also successfully applied in the analysis of bacterial culture medium with a detection limit of 0.64 μg/mL for bisphenol. The feeding experiments revealed that *L. lactis* is capable to uptake of bisphenol and contribute to its elimination from the culture medium. Further experiments are needed for the detection and quantification of metabolites as a function of culture time and administrated bisphenol concentration.

Supplementary Materials: The following are available online at http://www.mdpi.com/2297-8739/5/1/12/s1, Figure S1: Blank chromatograms of *L. lactis* cells (A) and cell free bacterial medium (B).

Acknowledgments: No additional funding was received in support of the present research work.

Author Contributions: M.T. conceived the experiments; M.T. and V.F.S. designed the experiments; A.T.R. performed the experiments; A.T.R. analyzed the data, A.T.R., M.T. and V.F.S. wrote the paper.

Conflicts of Interest: The authors declare no conflict of interest.

References

1. López-Cervantes, J.; Sánchez-Machado, D.; Paseiro-Losada, P.; Simal-Lozano, J. Effects Of Compression, Stacking, Vacuum Packing and Temperature on the Migration of Bisphenol A from Polyvinyl Chloride Packaging Sheeting into Food Simulants. *Chromatographia* **2003**, *58*, 327–330.
2. Kang, J.-H.; Kondo, F.; Katayama, Y. Human exposure to bisphenol A. *Toxicology* **2006**, *226*, 79–89. [CrossRef] [PubMed]
3. Ballesteros-Gómez, A.; Rubio, S.; Pérez-Bendito, D. Analytical methods for the determination of bisphenol A in food. *J. Chromatogr. A* **2009**, *16*, 449–469. [CrossRef] [PubMed]
4. Liu, J.; Martin, J.W. Prolonged exposure to bisphenol A from single dermal contact events. *Environ. Sci. Technol.* **2017**, *51*, 9940–9949. [CrossRef] [PubMed]
5. Völkel, W.; Colnot, T.; Csanady, G.; Filser, J.G.; Dekant, W. Metabolism and kinetics of bisphenol A in humans at low doses following oral administration. *Chem. Res. Toxicol.* **2002**, *15*, 1281–1287. [CrossRef] [PubMed]
6. Chouhan, S.; Yadav, S.; Prakash, K.; Swati, J.; Singh, S.P. Singh Effect of bisphenol A on human health and its degradation by microorganisms: A review. *Ann. Microbiol.* **2014**, *64*, 13–21. [CrossRef]
7. Kang, J.H.; Kondo, F. Bisphenol A Degradation by Bacteria Isolated from River Water. *Arch. Environ. Contam. Toxicol.* **2002**, *43*, 265–269. [CrossRef] [PubMed]
8. Tsai, W.T. Human Health Risk on Environmental Exposure to Bisphenol-A: A Review. *J. Environ. Sci. Health* **2006**, *24*, 225–255. [CrossRef] [PubMed]
9. Oehlmann, J.; Oetken, M.; Schulte-Oehlmann, U. A critical evaluation of the environmental risk assessment for plasticizers in the freshwater environment in Europe, with special emphasis on bisphenol A and endocrine disruption. *Environ. Res.* **2008**, *108*, 140–149. [CrossRef] [PubMed]
10. Staples, C.A.; Dorn, P.B.; Klecka, G.M.; O'Block, S.T.; Harris, L.R. A review of the environmental fate, effects, and exposures of bisphenol-A. *Chemosphere* **1998**, *36*, 2149–2173. [CrossRef]
11. Da Silva, J.C.C.; Reis Teodoro, J.A.; Afonso, R.J.; Aquino, S.F.; Augusti, R. Photodegradation of bisphenol A in aqueous medium: Monitoring and identification of by-products by liquid chromatography coupled to high-resolution mass spectrometry. *Rapid Commun. Mass Spectrom.* **2014**, *28*, 987–994. [CrossRef] [PubMed]
12. Diepens, M.; Gijsman, P. Photodegradation of bisphenol A polycarbonate. *Polym. Degrad. Stab.* **2007**, *92*, 397–406. [CrossRef]
13. Suzuki, T.; Nakagawa, Y.; Takano, I.; Yaguchi, K.; Yasuda, K. Environmental Fate of Bisphenol A and Its Biological Metabolites in River Water and Their Xeno-estrogenic Activity. *Environ. Sci. Technol.* **2004**, *38*, 2389–2396. [CrossRef] [PubMed]
14. Zhang, W.; Yin, K.; Chen, L. Bacteria-mediated bisphenol A degradation. *Appl. Microbiol. Biotechnol.* **2013**, *97*, 5681–5689. [CrossRef] [PubMed]
15. Kang, J.H.; Katayama, Y.; Kondo, F. Biodegradation or metabolism of bisphenol A: From microorganisms to mammals. *Toxicology* **2006**, *217*, 81–90. [CrossRef] [PubMed]
16. Kolvenbach, B.A.; Helbling, D.E.; Kohler, H.P.E.; Corvini, P.F.X. Emerging chemicals and the evolution of biodegradation capacities and pathways in bacteria. *Curr. Opin. Biotechnol.* **2014**, *27*, 8–14. [CrossRef] [PubMed]
17. Giarma, E.; Amanetidou, E.; Toufexi, A.; Touraki, M. Defense systems in developing *Artemia franciscana* nauplii and their modulation by probiotic bacteria offer protection against a *Vibrio anguillarum* challenge. *Fish Shellfish Immunol.* **2017**, *66*, 163–172. [CrossRef] [PubMed]
18. Oishi, K.; Sato, T.; Yokoi, W.; Yoshida, Y.; Ito, M.; Sawada, H. Effect of probiotics, *Bifidobacterium breve* and *Lactobacillus casei*, on bisphenol A exposure in rats. *Biosci. Biotechnol. Biochem.* **2008**, *72*, 1409–1415. [CrossRef] [PubMed]
19. Monachese, M.; Burton, J.P.; Reid, G. Bioremediation and tolerance of humans to heavy metals through microbial processes: A potential role for probiotics? *Appl. Environ. Microbiol.* **2012**, *78*, 6397–6404. [CrossRef] [PubMed]
20. Yamanaka, H.; Moriyoshi, K.; Ohmoto, T.; Ohe, T.; Sakai, K. Degradation of bisphenol A by *Bacillus pumilus* isolated from kimchi, a traditionally fermented food. *Appl. Biochem. Biotechnol.* **2007**, *136*, 39–51. [CrossRef] [PubMed]
21. Vijayalakshmi, G.; Ramadas, V.; Nellaiah, H. Isolation, indentification and degradation of Biphenol A by *Bacillus* sp. from effluents of thermal paper industry. *Int. J. Sci. Eng. Res.* **2013**, *4*, 366–374.

22. Zühlke, M.K.; Schlüter, R.; Henning, A.K.; Lipka, M.; Mikolasch, A.; Schumann, P.; Giersberg, M.; Kunze, G.; Schauer, F. A novel mechanism of conjugate formation of bisphenol A and its analogues by *Bacillus amyloliquefaciens*: Detoxification and reduction of estrogenicity of bisphenols. *Int. Biodeterior. Biodegrad.* **2016**, *109*, 165–173. [CrossRef]
23. Endo, Y.; Kimura, N.; Ikeda, I.; Fujimoto, K.; Kimoto, H. Adsorption of bisphenol A by lactic acid bacteria, *Lactococcus*, strains. *Appl. Microbiol. Biotechnol.* **2007**, *74*, 202–207. [CrossRef] [PubMed]
24. Fujiwara, H.; Soda, S.; Fujita, M.; Ike, M. Kinetics of bisphenol A degradation by *Sphingomonas paucimobilis* FJ-4. *J. Biosci. Bioeng.* **2016**, *122*, 341–344. [CrossRef] [PubMed]
25. Im, J.; Yip, D.; Lee, J.; Löffler, F.E. Simplified extraction of bisphenols from bacterial culture suspensions and solid matrices. *J. Microbiol. Methods* **2016**, *126*, 35–37. [CrossRef] [PubMed]
26. Spivack, J.; Leib, T.K.; Lobos, J.H. Novel Pathway for Bacterial Metabolism of Bisphenol A. *J. Biol. Chem.* **1994**, *269*, 7323–7329. [PubMed]
27. Lobos, J.H.; Leib, T.K.; Su, T.M. Biodegradation of bisphenol A and other bisphenols by a gram-negative aerobic bacterium. *Appl. Environ. Microbiol.* **1992**, *58*, 1823–1831. [PubMed]
28. Rykowska, I.; Wasiak, W. Properties, threats, and methods of analysis of bisphenol A and its derivatives. *Acta Chromatogr.* **2006**, *16*, 7–27.
29. Xiao, Q.; Li, Y.; Ouyang, H.; Xu, P.; Wu, D. High-performance liquid chromatographic analysis of bisphenol A and 4-nonylphenol in serum, liver and testis tissues after oral administration to rats and its application to toxicokinetic study. *J. Chromatogr. B* **2006**, *830*, 322–329. [CrossRef] [PubMed]
30. Samanidou, V.F.; Frysali, M.A.; Papadoyannis, I.N. Matrix solid phase dispersion for the extraction of bisphenol A from human breast milk prior to HPLC analysis. *J. Liquid Chromatogr. Relat. Technol.* **2014**, *37*, 247–258. [CrossRef]
31. Samanidou, V.F.; Kerezoudi, C.; Tolika, E.; Palaghias, G. A Simple Isocratic HPLC Method for the Simultaneous Determination of the Five Most Common Residual Monomers Released from Resin-Based Dental Restorative Materials. *J. Liquid Chromatogr. Relat. Technol.* **2015**, *38*, 740–749. [CrossRef]
32. Samanidou, V.; Hadjicharalampous, M.; Palaghias, G.; Papadoyannis, I. Development and validation of an isocratic HPLC method for the simultaneous determination of residual monomers released from dental polymeric materials in artificial saliva. *J. Liquid Chromatogr. Relat. Technol.* **2012**, *35*, 511–523. [CrossRef]
33. Asimakopoulos, A.G.; Thomaidis, N.S.; Koupparis, M.A. Recent trends in biomonitoring of bisphenol A, 4-*t*-octylphenol, and 4-nonylphenol. *Toxicol. Lett.* **2012**, *210*, 141–154. [CrossRef] [PubMed]
34. Chen, G.W.; Ding, W.H.; Ku, H.Y.; Chao, H.R.; Chen, H.Y.; Huang, M.C.; Wang, S.L. Alkylphenols in human milk and their relations to dietary habits in central Taiwan. *Food Chem. Toxicol.* **2010**, *48*, 1939–1944. [CrossRef] [PubMed]
35. Luo, L.; Yang, Y.; Wang, Q.; Li, H.P.; Luo, Z.F.; Qu, Z.P.; Yang, Z.G. Determination of 4-*n*-octylphenol, 4-*n*-nonylphenol and bisphenol A in fish samples from lake and rivers within Hunan Province China. *Microchem. J.* **2017**, *132*, 100–106. [CrossRef]
36. Neamtu, M.; Frimmel, F.H. Degradation of endocrine disrupting bisphenol A by 254 nm irradiation in different water matrices and effect on yeast cells. *Water Res.* **2006**, *40*, 3745–3750. [CrossRef] [PubMed]
37. Matsumura, Y.; Hosokawa, C.; Sasaki-Mori, M.; Akahira, A.; Fukunaga, K.; Ikeuchi, T.; Oshiman, K.; Tsuchido, T. Isolation and characterization of novel bisphenol-A-degrading bacteria from soils. *Biocontrol Sci.* **2009**, *14*, 161–169. [CrossRef] [PubMed]
38. Beltran-Heredia, J.; Torregrosa, J.; Dominguez, J.R.; Peres, J.A. Comparison of the degradation of p-hydroxybenzoic acid in aqueous solution by several oxidation processes. *Chemosphere* **2001**, *42*, 351–359. [CrossRef]
39. Abbas, S.S.; Elghobashy, M.R.; Bebawy, L.I.; Shokry, R.F. Stability-indicating chromatographic determination of hydroquinone in combination with tretinoin and fluocinolone acetonide in pharmaceutical formulations with a photodegradation kinetic study. *RSC Adv.* **2015**, *5*, 43178–43194. [CrossRef]
40. Eio, E.J.; Kawai, M.; Tsuchiya, K.; Yamamoto, S.; Toda, T. Biodegradation of bisphenol A by bacterial consortia. *Int. Biodeterior. Biodegrad.* **2014**, *96*, 166–173. [CrossRef]

Article

Silk Fibroin Nanoparticles for Drug Delivery: Effect of Bovine Serum Albumin and Magnetic Nanoparticles Addition on Drug Encapsulation and Release

Olga Gianak [1], Eleni Pavlidou [2], Charalambos Sarafidis [2], Vassilis Karageorgiou [3] and Eleni Deliyanni [4,*]

1 Division of Analytical Chemistry, Department of Chemistry, Aristotle University of Thessaloniki, 54124 Thessaloniki, Greece; gianakolga@hotmail.com
2 Department of Physics, Aristotle University of Thessaloniki, 54124 Thessaloniki, Greece; elpavlid@auth.gr (E.P.); hsara@auth.gr (C.S.)
3 Department of Food Technology, ATEI of Thessaloniki, P.O. BOX 141, 57400 Thessaloniki, Greece; vkarageorgiou@food.teithe.gr
4 Division of Chemical Technology, Department of Chemistry, Aristotle University of Thessaloniki, 54124 Thessaloniki, Greece
* Correspondence: lenadj@chem.auth.gr; Tel.: +30-2310-997808

Received: 4 February 2018; Accepted: 9 April 2018; Published: 23 April 2018

Abstract: Silk fibroin nanoparticles were prepared in the present study based on phase separation between silk fibroin and polyvinyl alcohol. The drug encapsulation efficiency of the prepared nanoparticles was examined at a range of concentrations from 10 ppm to 500 ppm for pramipexole, curcumin, and propranolol hydrochloride. Silk fibroin nanoparticles encapsulated with propranolol presented the highest drug release profile. In order to improve the drug encapsulation efficiency and drug release performance, a modification of silk fibroin nanoparticles with bovine serum albumin and magnetic nanoparticles was tried. The modification was found to improve the drug encapsulation and release of the modified nanoparticles. Bovine-serum-modified nanoparticles presented the best improvement.

Keywords: silk fibroin; drug delivery; magnetic silk fibroin; bovine serum albumin

1. Introduction

Drug delivery aims to bring the compound with pharmaceutical activity to the exact point of need in the organism with the right concentrations in order to increase the efficiency of action [1]. The improvement of cellular uptake, the reduction of the side effects, the control of drug release, the enhancement of drug bioavailability, and the reduction of the drug degradation rate are the purposes of drug delivery systems [2]. Recently, nanoparticles are considered to be suitable for drug delivery due to their ability to act as modifiable platforms, their tunable size, and their high surface-to-volume ratio [2]. In addition, they can be used to deliver hydrophilic and hydrophobic drug molecules as well [3].

Silk fibroin, from the silk worm Bombyx mori, consists of hydrophobic and hydrophilic regions [2]. The hydrophobic domains, i.e., protein crystals and beta sheets, are dominated by repeats of alanine, glycine-alanine, and glycine-alanine-serine. Recently, silk fibroin, due to its excellent properties, such as biocompatibility, biodegradability, and low immunogenicity, has been extensively tested for a drug delivery application since silk materials exhibit high encapsulation efficiency and controllable drug release kinetics [4–6].

Bovine serum albumin (BSA) is a water-soluble protein with a well-defined structure. Positively or negatively charged drug molecules can be non-covalently bound to the charged amino acids of albumin. Silk fibroin contains hydrophobic amino acids that could create strong electrostatic interactions through the amino acids' carboxyl groups and the amino groups of bovine serum albumin. These interactions prevent leakage of hydrophobic drugs from bovine serum albumin; thereby, it is expected to improve drug encapsulation, delivery to affected body areas, and drug release rate [7]. Albumin was found to enable the localization of drug carriers in specific tissues (i.e., liver and heart tissues), regulating in this way biodistribution and drug release [8]. Besides this, it was shown that readily available and inexpensive serum albumin can overcome many drugs' shortcomings, such as insolubility, instability in biological environments, poor uptake into cells and tissues, sub-optimal selectivity for targets, and unwanted side effects [9].

Additionally, magnetic carriers and external magnetic fields that focalize on tumors have emerged as a hopeful strategy to enhance drug accumulation at tumor sites. Magnetic targeting has the advantage of not requiring complex chemical modification of targeting ligands on the surface of nanocarriers when compared with conventional tumor targeting. Exploiting a magnetic field as a driving force represents a noninvasive therapeutic approach [10,11].

In the present study, we report aqueous-based preparation methods for silk fibroin nanoparticles, based on phase separation between silk fibroin and polyvinyl alcohol (PVA), with a simple, inexpensive, and appropriate method for pharmaceutical and biomedical applications of drug delivery. Curcumin, pramipexole, and propranolol hydrochloride were chosen as model drugs in this study for their different molecular weights (MWs), surface chemistries, and disease targets. Curcumin, a natural compound with diphenolic groups, is extracted from the rhizome of turmeric. It has been used in clinical trials for cancer therapy [12,13]. Propranolol is a non-selective beta adrenergic blocking agent and has been used in the treatment of hypertension, angina pectoris, and many other cardiovascular disorders [14]. Pramipexole is a potent dopamine D_2 agonist with a preference for D_3 receptors and is approved worldwide for the treatment of Parkinson's disease symptoms [15].

Moreover, nanoparticles of silk fibroin were modified with bovine serum albumin for the enrichment of silk fibroin with the above-mentioned properties of albumin [7–9]. Silk fibroin-bovine serum albumin nanoparticles were prepared since bovine serum albumin has been found to be biocompatible, biodegradable, non-toxic, and non-immunogenic and has interesting results in drug encapsulation and release [16]. Finally, magnetic silk fibroin nanoparticles were also prepared and examined to identify the contribution of magnetic nanoparticles to the encapsulation efficiency and in vitro drug release of the model drugs.

2. Materials and Methods

2.1. Materials

Polyvinyl alcohol high molecular weight solid, (PVA 98-99 hydrolized) and curcumin (95% total curcuminoid content) from Tumeric rhizome were purchased from A Johnson Company (New Brunswick, NJ, USA). Ethanol, Lithium Bromide 99%, and Bovine Serum Albumin (BSA) (lyophilized powder, 66.000 kDa), were purchased from Sigma-Aldrich (St. Louis, MO, USA). Propranolol hydrochloride and Pramipexole were purchased from Fagron Hellas (Trikala, Greece). $FeCl_3{\bullet}6H_2O$, $FeCl_2{\bullet}4H_2O$, and NH_4OH of analytical grade were purchased from Sigma-Aldrich.

2.2. Methods

2.2.1. Preparation of Silk Fibroin Solution

Silk fibroin aqueous solution was prepared according to described protocols [17]. Briefly, 5 g cocoons of Bombyx mori were boiled in 2 L of an aqueous solution of sodium carbonate (0.02 M) for 30 min and the degummed silk fibroin was thoroughly rinsed with deionized water, air dried overnight, and then dissolved in a 9.3 M LiBr solution at 60 °C. The solution was dialyzed against deionized water using Slide-a-Lyzer dialysis cassettes (MWCO 3.500, Pierce, Thermo-Fischer Scientific, Waltham, MA, USA) for 2 days for salt removal. In order to clean the solution from impurities, silk aggregates, and debris of cocoons, the silk fibroin solution was centrifuged twice at 9000 rpm for 30 min. The prepared silk fibroin solution was stored at 4 °C for further use.

2.2.2. Preparation of Silk Fibroin (SF) Nanoparticles

Silk fibroin particles were prepared according to the following method [18]: 5 mL of silk fibroin solution was mixed with 2 mL ethanol with a sample pipette and then vortexed for 10 s. Subsequently, 50 mL of 5% PVA solution were added to the fibroin/ethanol mixture and vortexed for 10 s. The fibroin/ethanol/PVA solution mixture was refrigerated for 24 h and centrifuged. In order to remove PVA and ethanol from the nanoparticles, the mixture was washed with deionized water three times and centrifuged after each wash [18].

2.2.3. Preparation of Silk Fibroin-Bovine Serum Albumin (SF-BSA) Nanoparticles

For the preparation of the Silk Fibroin-Bovine Serum Albumin nanoparticles (SF-BSA-NPs), 2.5 mL of the silk fibroin solution were mixed with 2.5 mL of 2% BSA solution and then the mixture was stirred for 15–20 min. The SF-BSA-NPs were prepared following the method used to prepare the silk fibroin nanoparticles.

2.2.4. Synthesis of Fe_3O_4 Nanoparticles

For the preparation of the magnetic silk fibroin nanoparticles (SFm-NPs), the magnetic material used was magnetite nanoparticles (Fe_3O_4) prepared in the laboratory. The Fe_3O_4 nanoparticles were prepared according to the modified Massart method [19] via the co-precipitation of $FeCl_3{\bullet}6H_2O$ and $FeCl_2{\bullet}4H_2O$ [18]; briefly, 11.2 mmol $FeCl_3{\bullet}6H_2O$ and 5.6 mmol $FeCl_2{\bullet}4H_2O$ were dissolved in 150 mL of deionized water, heated to 60 °C under agitation in an inert atmosphere, and then aqueous ammonia solution was added dropwise until the solution's pH became 10. The precipitate that was formed was magnetically collected, washed with deionized water and ethanol, and freeze-dried.

2.2.5. Preparation of Silk Fibroin-Fe_3O_4 (SFm) Nanoparticles

For the preparation of the magnetic silk fibroin nanoparticles (SFm-NPs), 0.1 g Fe_3O_4 was added to 5 mL of silk fibroin solution followed by the preparation of nanoparticles as above described.

2.2.6. Preparation of Drug-Encapsulated Silk Fibroin Nanoparticles

Curcumin (MW = 368.39 g/mol), propranolol hydrochloride (MW = 259.34 g/mol), and pramipexole (MW = 211.32 g/mol) were the model drugs tested for the encapsulation in the silk fibroin nanoparticles; the chemical formulas of the drugs are presented in Table 1.

Amounts of the pramipexole and propranolol were dissolved in deionized water and mixed with 5 mL of the silk fibroin solution; curcumin was dissolved in ethanol and then mixed with 5 mL of the silk fibroin solution. Different amounts of the drugs were estimated for the preparation of solutions with concentrations of 10, 35, 50, 100, 200, and 500 ppm for each drug. Finally, the drug-encapsulated silk fibroin nanoparticles were prepared as previously described.

The encapsulation efficiency in the silk fibroin nanoparticles was determined by measuring the UV-vis absorbance (HITACHI U-2000) of the supernatant after the centrifugation that follows the formation of silk fibroin nanoparticles at 430, 265, and 235 nm for curcumin, pramipexole, and propranolol, respectively. Standard calibration curves of model drugs were used for drug quantification ($r^2 \geq 0.999$). All experiments were performed in triplicate and the mean value is presented. The difference between the total amount of drug used in the experiment and the amount that remained in the supernatants was expressed as encapsulation efficiency (EE) in silk fibroin nanoparticles [20] and determined by the following equation:

$$\text{Encapsulation Efficiency} = \frac{\text{amount of drug that remained in the particles}}{\text{total amount of drug used}} \times 100 \quad (1)$$

Table 1. Model drugs encapsulated in silk fibroin nanoparticles.

Drug	Formula	MW
curcumin	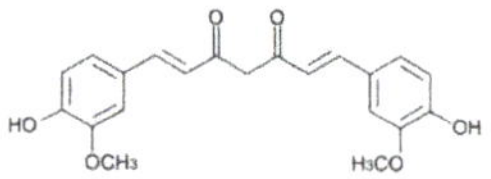 HO, OCH₃, H₃CO, OH	368.39
propranolol hydrochloride	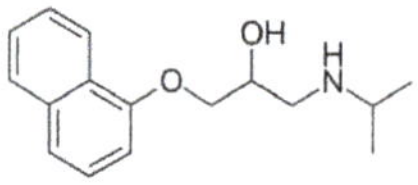 OH, O, H, N	259.34
pramipexole	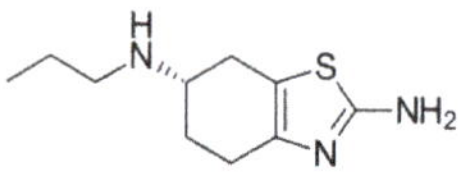 H, N, S, NH₂, N	211.32

MW = molecular weight.

2.2.7. Silk Fibroin Nanoparticles Characterization

For the identification of the crystalline phase of silk fibroin nanoparticles, X-ray powder diffraction (XRD) patterns were recorded on a Philips PW 1820 diffractometer (Amsterdam, The Netherlands) with Cu Kα radiation from 20° to 60°. The morphology of silk fibroin nanoparticles as well as that of drug-encapsulated silk fibroin nanoparticles was imaged using a JEOL JMS-840A scanning electron microscope (JEOL, Tokyo, Japan). The control samples of silk fibroin nanoparticles suspensions in water, the samples of SF-BSA-NPs and SFm-NPs, and their respective drug-encapsulated nanoparticles suspensions in water were lyophilized and the powder obtained was subjected to Fourier Transform Infrared measurement using a Perkin-Elmer FTIR spectrophotometer (model Spectrum 1000, Rodgau, Germany). Nanoparticle size, size distribution, and surface charges of the nanoparticles, and those of their respective drug-encapsulated nanoparticles, were determined by Dynamic Light Scattering (DLS) with a NanoBrook ZetaPALS Brookhaven Instruments (Holtsville, NY, USA) equipped with a diode laser (wavelength, λ = 532 nm). Hysteresis loops for the SF-magnetite nanoparticles were recorded with an Oxford 1.2 H/CF/HT vibrating sample magnetometer (VSM).

2.2.8. In Vitro Drug Release

The amount of drug released in vitro from the drug encapsulated silk fibroin nanoparticles, at specific time intervals, was calculated as follows: 4 mL of the drug-encapsulated nanoparticles were added to 20 mL of phosphate buffer saline (PBS) at pH 7.4 and shaken at a rate of 120 rpm at 37 °C. At specified time intervals, 2 mL were sampled and centrifuged at 9000 rpm for 10 min. The concentration of the drug in the supernatant was measured by UV-vis spectrometry at 430, 265, and 235 nm for curcumin, pramipexole, and propranolol, respectively. The percentage release was determined as the ratio of the measured amount of the drug released at different time intervals to the initial amount of drug encapsulated in the nanoparticles [12]. All experiments were performed in triplicate and the mean value is presented.

3. Results and Discussion

Weighting the residual solid of a certain known volume of silk solution after drying at 60 °C, the concentration of the silk fibroin aqueous solution prepared was found to be approximately 4.6% (*w*/*v*).

3.1. XRD Characterization, Morphology, and FTIR Characterization of the Silk Fibroin Nanoparticles

XRD peaks of silk fibroin are associated with its crystalline structure; Figure 1 presents the XRD results of the silk fibroin nanoparticles prepared in the current study. The 2θ diffraction peak presented in the XRD pattern of the silk fibroin nanoparticles at 2θ = 19.2° could be attributed to the silk II crystal structure of fibroin, indicating an increased crystallization degree of silk fibroin in the nanoparticles [21,22].

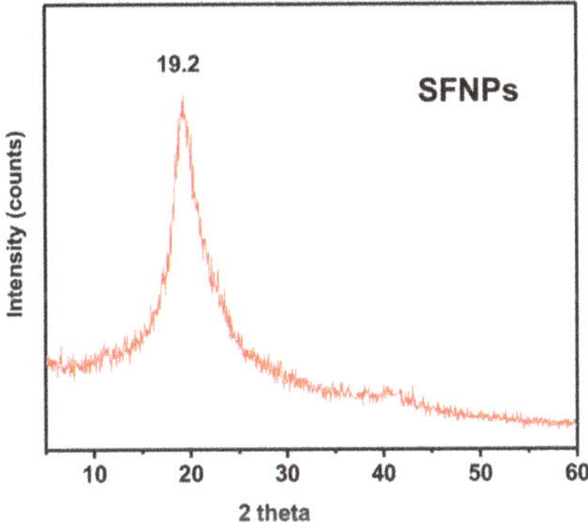

Figure 1. XRD patterns of silk fibroin nanoparticles (SFNPs).

The morphology of silk fibroin nanoparticles was illustrated by SEM images and is presented in Figure 2a. The nanoparticles observed presented a spherical and/or cylindrical shape with an average mean diameter of about 290 nm. Silk fibroin nanoparticles previously prepared with the same method by Shi et al. [18] presented an average diameter of about 686.5 nm as estimated by SEM image. This difference in size could be attributed to the lower concentration (4%) of silk fibroin solution that was used in this study to prepare the silk fibroin nanoparticles compared to the study by Shi et al. (6%). According to Zhao et al., the increase of the starting silk fibroin solution concentration has an effect on the particle size increase [23].

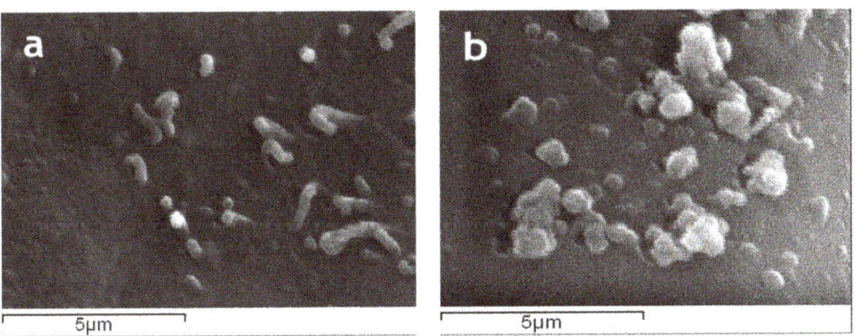

Figure 2. *Cont.*

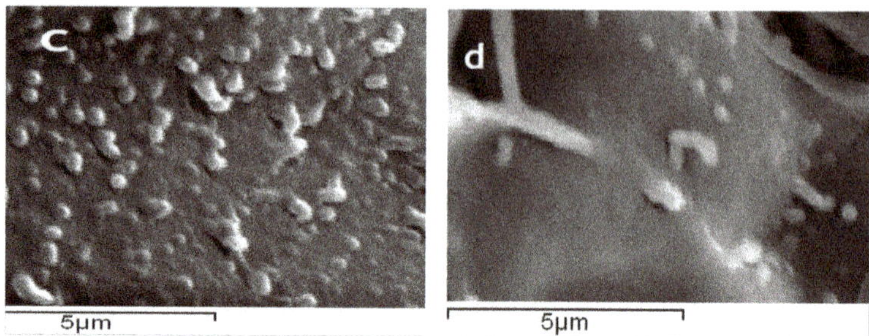

Figure 2. SEM images of SF nanoparticles (**a**); SF-propanolol NPs (**b**); SF-pramipexole NPs (**c**); and SF-curcumin NPs (**d**).

In Figure 3, the FTIR spectrum of the prepared silk fibroin nanoparticles is presented. The characteristic peak at 1632 cm^{-1} attributed to an amide I β-sheet and the peak at 1260 cm^{-1} attributed to amide III are due to random coil. So, when the nanoparticles of silk fibroin were formed both the silk I crystal structure and the silk II crystal structure of silk fibroin were observed [24,25]. The assignments of the FTIR bands of silk fibroin nanoparticles are presented in detail in Table 2.

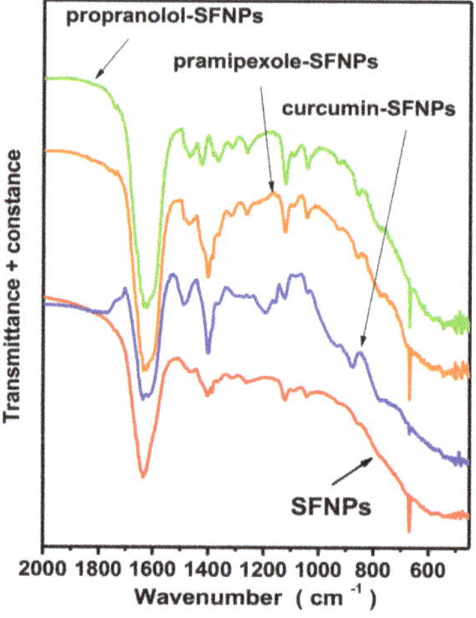

Figure 3. FTIR spectra of silk fibroin nanoparticles and drug-encapsulated silk fibroin nanoparticles.

Table 2. Assignments of the FTIR bands of silk fibroin nanoparticles and drug-encapsulated silk fibroin nanoparticles.

cm^{-1}	SF Nanoparticles	cm^{-1}	SF-Pramipexole NPs	cm^{-1}	SF-Propranolol NPs	cm^{-1}	SF-Curcumin NPs
1632	Amide I β-sheets	1627	Amide I β-sheets	1627	Amide I β-sheets	1628	Amide I β-sheets
1465	CH_3 bending	1469	CH_3 bending	1465	CH_3 bending	1488	curcumin
1401	CH_3 bending			1422	CH_3 bending	1399	CH_3 bending
1354	CH_2 bending $(AG)_n$			1367	CH_2 bending $(AG)_n$		
1311	OCO– stretching	1315	OCO– stretching	1319	OCO– stretching	1303	Amide III β-turn
1260	Amide III a-helices	1260	Amide III a-helices	1260	Amide III a-helices		
						1192	curcumin
1123	CH_2OH polyphenols	1118	CH_2OH polyphenols	1118	CH_2OH polyphenols		
1041	C–O stretching C–OH stretching Ser	1041	C–O stretching C–OH stretching Ser	1041	C–O stretching C–OH stretching Ser	1036	C–O stretching C–OH stretching Ser
926	$(A)_n$ β-sheets	926	$(A)_n$ β-sheets	926	$(A)_n$ β-sheets		
						873	curcumin

$(A)_n$, polyalanine; $(AG)_n$, polyalanine glycine [26].

3.2. Encapsulation Efficiency

The encapsulation efficiency (EE) of the silk fibroin nanoparticles for the three model drugs was studied at a range of concentrations from 10 to 500 ppm. For the curcumin-encapsulated silk fibroin nanoparticles the highest EE was found to be 98 ± 2% and was achieved at 500 ppm, for propranolol the highest EE was 69 ± 3% at 50 ppm and for pramipexole the highest EE was 68 ± 2% at 35 ppm (data not shown).

3.3. Size, ζ-Potential, Morphology, and FTIR Characterization of Silk Fibroin Nanoparticles and Drug-Encapsulated Silk Fibroin Nanoparticles

The size of silk fibroin nanoparticles measured by DLS is presented in Table 3. As seen in the Table, the average diameter of the silk fibroin nanoparticles was found to be 301 ± 11 nm, which is consistent with the size estimated by SEM; the average size was found to increase after the encapsulation of the model drugs (curcumin, pramipexole, and propranolol) in the silk fibroin nanoparticles. The polydispersity index, ranging from 0.10 to 0.25, revealed a uniform particle distribution.

Table 3. Diameter of silk fibroin nanoparticles measured by dynamic light scattering.

	Diameter ± SD (nm) (n = 5)	Polydispersity Index
SFNPs	301 ± 11	0.24
SF-curcumin NPs	354 ± 13	0.25
SF-pramipexole NPs	469 ± 14	0.10
SF-propranolol NPs	501 ± 7	0.20

The surface charge of the silk fibroin solution, of the drug solutions, of the silk fibroin nanoparticles, and of the silk fibroin nanoparticles encapsulated with the model drugs was determined by ζ-potential and the results are presented in Table 4. For all samples, the ζ-potential was negative; it was observed that the ζ-potential of propranolol-encapsulated silk fibroin nanoparticles (SF-propranolol NPs) was higher than that of curcumin (SF-curcumin NPs) and pramipexole-encapsulated silk fibroin nanoparticles (SF-pramipexole NPs), leading to the conclusion that propranolol was encapsulated in the center of the silk fibroin nanoparticles while curcumin and pramipexol were adsorbed on the surface of the nanoparticles.

Table 4. ζ-potential of model drug solutions, silk fibroin nanoparticles, and drug-encapsulated silk fibroin nanoparticles.

Sample	ζ-Potential ± SD (mV) (n = 3)
Fibroin solution 4.1%	−4.3 ± 0.1
Curcumin solution 500 ppm	−7 ± 2
Pramipexole solution 35 ppm	−0.06 ± 0.02
Propranolol solution 50 ppm	−2.1 ± 0.3
Silk fibroin nanoparticles	−16.40 ± 0.02
SF-curcumin NPs	−5.5 ± 0.2
SF-pramipexole NPs	−6 ± 2
SF-propranolol NPs	−14.9 ± 0.3

The morphology of drug-encapsulated silk nanoparticles, and that of the silk nanoparticles for the sake of comparison, was illustrated by SEM images (Figure 2). The nanoparticles observed presented a spherical and/or cylindrical shape with an average mean diameter of 290 nm. As expected, the size of silk fibroin nanoparticles increased after drug encapsulation. The mean diameter of the drug-encapsulated silk fibroin nanoparticles, as estimated from the SEM images, was 490, 392, and 294 nm for propranolol-, pramipexole-, and curcumin-encapsulated silk fibroin nanoparticles, respectively. It is worth noting that these results for nanoparticle sizes are relatively smaller than those measured with DLS; this can be attributed to the dry state of the nanoparticles when measured with SEM.

In Figure 3, the FTIR spectra of silk fibroin nanoparticles and of drug-encapsulated silk fibroin nanoparticles are presented. In the spectra of the curcumin-encapsulated silk fibroin nanoparticles, it was observed that certain characteristic peaks of the original curcumin were either not evident or exhibited a slight shift or alteration, which demonstrated that certain structural changes and chemical reactions may have occurred between the curcumin and the silk fibroin [12,27]. Curcumin showed its characteristic group absorption peaks in sharp absorption bands at 1605 cm^{-1}, 1502 cm^{-1} (–C=O and –C–C vibrations), 1435 cm^{-1} (an olefinic –C–H bending vibration), 1285 cm^{-1} (an aromatic –C–O stretching vibration), and 833 cm^{-1} (a C–H bond of alkene group) [12,27]. In addition, propranolol exhibited characteristic peaks in sharp absorption bands at 1680–1620 cm^{-1} (C=C stretching) and 1260–1000 cm^{-1} (C–O stretching) [28]. On the contrary, the FTIR spectrum of pramipexole-encapsulated fibroin nanoparticles presented no difference compared to pure fibroin nanoparticles; the characteristic peaks of pramipexol were not evident, leading to the conclusion that after encapsulation with pramipexole the chemical bonding remained unchanged [29]. The assignments of the FTIR bands of silk fibroin nanoparticles as well as those of the drug-encapsulated silk fibroin nanoparticles are presented in detail in Table 4.

3.4. In Vitro Drug Release

The drug release profile, presented in Figure 4, indicated that the encapsulated curcumin and pramipexole silk fibroin nanoparticles had a short and low-level release percentage, about 1.2% and 0.25%, respectively (presented also in the insets of the Figure 4) while the propranolol-encapsulated silk fibroin nanoparticles exhibited a release of about 65% within 6 days. This could be due to the lack of interactions of propranolol with the silk fibroin as well as to the hydrophilic properties of propranolol. Curcumin, as a hydrophobic drug, was possibly attached to silk fibroin via hydrophobic interactions and π-π stacking, and presented a reduced release rate compared with that of the hydrophilic drug propranolol [4,12,30]. Pramipexole was expected to be attached to silk fibroin through strong electrostatic interactions attributed to the $–NH_2$ groups, which could have resulted in the decreased release rate.

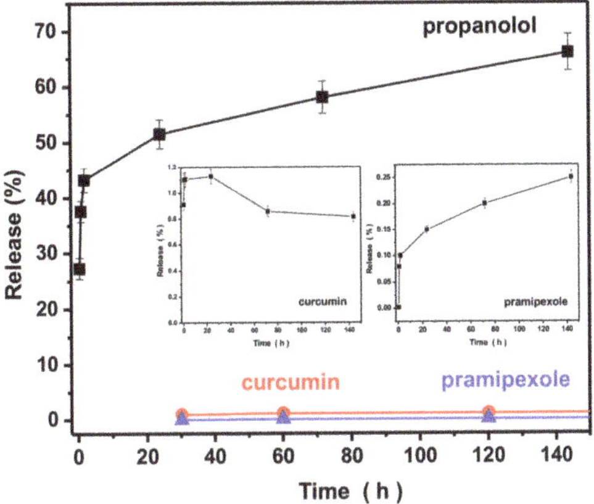

Figure 4. Drug release rate of model drugs from nanoparticles of silk fibroin.

3.5. *Modification of SF with Bovine Serum Albumin and Magnetic Nanoparticles*

In order to enhance the properties of silk fibroin nanoparticles, SFm-NPs and silk fibroin–bovine serum albumin nanoparticles (SF-BSA-NPs) were prepared via the introduction of magnetic nanoparticles (Fe_3O_4) and BSA during the silk fibroin nanoparticle formulation, and the drug encapsulation and release efficiency of these silk fibroin nanoparticles were examined in detail. Since propranolol presented the highest encapsulation and release performance, it was used as the model drug for the performance of the two modified types of silk fibroin nanoparticles.

3.5.1. Characterization of Modified Nanoparticles

DLS measurements for silk fibroin-bovine serum albumin nanoparticles and magnetic silk fibroin nanoparticles are presented in Table 5. Both samples indicated a small polydispersity index of 0.30 and 0.36, respectively, indicating a uniform diameter distribution of nanoparticles.

Table 5. Diameter of magnetic and fibroin-albumin nanoparticles measured by Dynamic Light Scattering.

Sample	Diameter ± SD (nm) (n = 5)	Polydispersity Index
SFNPs	301 ± 11	0.24
SFm-NPs	337 ± 14	0.30
SF-BSA-NPs	268 ± 4	0.36

SFm-NPs = magnetic silk fibroin nanoparticles; SF-BSA-NPs = silk fibroin-bovine serum albumin nanoparticles.

The encapsulation efficiency of magnetic silk fibroin and silk fibroin–bovine serum albumin nanoparticles is presented in Table 6. From the Table, it can be seen that the modification of silk fibroin with magnetic nanoparticles and bovine serum albumin increased the drug encapsulation compared to the pure silk fibroin nanoparticles. Propranolol is a hydrophilic drug and the presence of bovine serum albumin makes the silk fibroin–bovine serum albumin nanoparticles more hydrophilic; for this reason the encapsulation efficiency was increased. Also, the increased encapsulation efficiency of magnetic silk fibroin loaded with propranolol may be attributed to the adsorption of propranolol on Fe_3O_4.

Table 6. Encapsulation Efficiency (EE) of magnetic silk fibroin and silk fibroin-bovine serum albumin nanoparticles encapsulated with propranolol.

C (ppm)	EE ± SD (%) Propranolol SFNPs (n = 3)	EE ± SD (%) Propranolol SFm-NPs (n = 3)	EE ± SD (%) Propranolol SF-BSA-NPs (n = 3)
50	70 ± 2	96 ± 4	97 ± 3

The surface morphology of magnetic silk fibroin particles and the silk fibroin-bovine serum albumin nanoparticles as well as their encapsulated-with-propranolol counterparts, illustrated by SEM images, is presented in Figure 5. The magnetic silk fibroin (SFm) and silk fibroin-bovine serum albumin nanoparticles (SF-BSA) exhibited a spherical shape with a diameter of about 288 nm, while the propranolol-encapsulated silk fibroin-bovine serum albumin nanoparticles (SF-BSA + propranolol) and propranolol-encapsulated magnetic silk fibroin nanoparticles (SFm + propranolol) exhibited a particle size of about 384 nm and 490 nm, respectively (Figure 5).

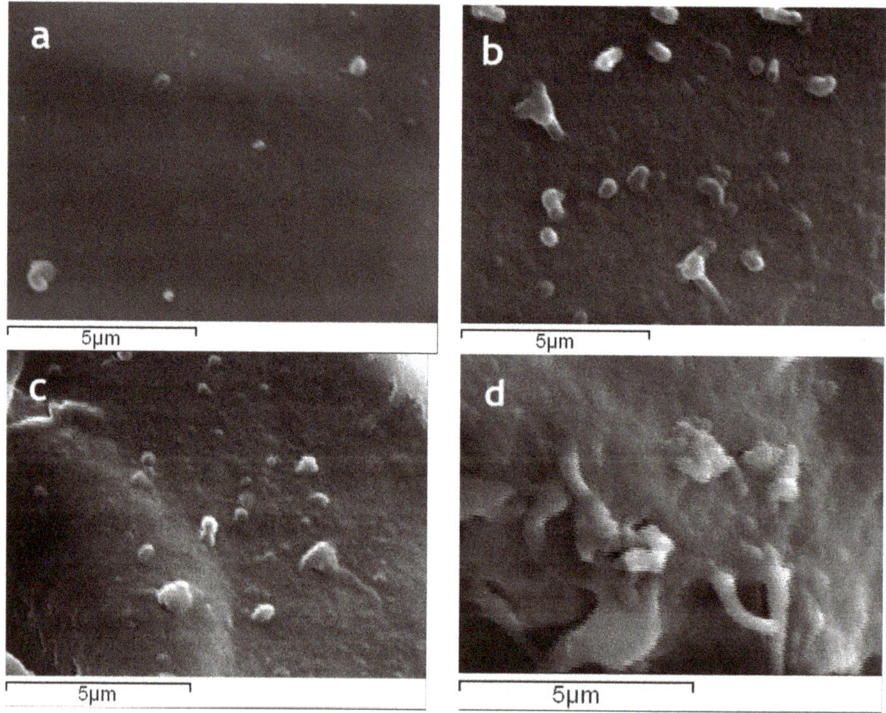

Figure 5. SEM images of SFm nanoparticles (**a**); SFm + propranolol NPs (**b**); SF-BSA-NPs (**c**); and SF-BSA + propranolol NPs (**d**).

The 2θ diffraction peak presented in the XRD pattern of the magnetic silk fibroin nanoparticles (Figure 6a), that appeared at 2θ = 19.2°, could be attributed to silk I, a random coil of silk fibroin, while for the encapsulated-with-propranolol magnetic silk fibroin nanoparticles, the characteristic peak at 2θ = 20.7° could be attributed to silk II, a β-sheet of fibroin. In addition, diffraction peaks corresponding to the (3 1 1), (4 0 0), and (4 4 0) planes of the cubic crystal structure (fcc) of Fe_3O_4, show the formation of magnetic silk fibroin nanoparticles [19]. The aforementioned XRD results are in agreement with the FTIR results.

Prior to the measurements, the instrument was calibrated against a NIST-certified Ni standard. Each sample had approximately the same dimensions and an additional calibration procedure was performed in order to minimize shape effects. All measurements were performed in a controlled room temperature of 22 °C and consisted of a gradual increase of the applied external magnetic field up to about +1.9 T (μ_0H) then down to −1.9 T and again up to +1.9 T while measuring the magnetic moment of the sample. The magnetic properties are attributed to the Fe_3O_4 particles and the overall appearance of the loop is typical for nanoparticles of that composition and size. All measurements have exactly the same appearance, i.e., initial slope, curvature, and saturation plateau, indicating that there are no significant differences in the Fe_3O_4 particles between the samples and that the procedure for the preparation of the magnetic silk fibroin nanoparticles (SFm-NPs) does not affect significantly the magnetic properties of the Fe_3O_4 particles. Remanence, Mr, the Mass Magnetization which corresponds to a zero field after the first application of the maximum external field equals to 5–7% of the saturation magnetization and coercivity, μ_0H_c, the reverse external field which zeroes the magnetic moment of the sample is about 0.01 T. Both the remanence and the coercivity are very low in all cases as expected for a soft ferromagnetic material. Saturation magnetization, the maximum Mass Magnetization which could be observed, was obtained for the sample via extrapolation from data of Figure 6b,c and was found to be σ_s = 27.0 (8b) and 27.9 (8c) emu/g. These results correspond to about 50% wt. Fe_3O_4 content, since the saturation magnetization depends on the Fe_3O_4 content which is diluted in the silk fibroin, confirmed that Fe_3O_4 particles are successfully incorporated into the silk fibroin and that their magnetic properties are preserved after drug encapsulation.

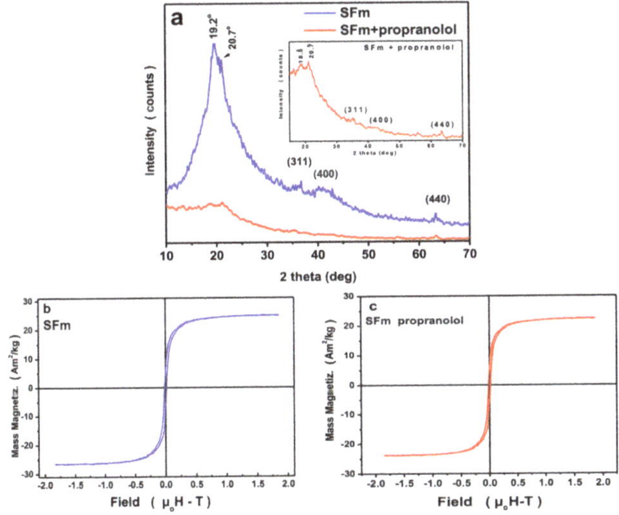

Figure 6. (**a**) XRD patterns of the magnetic silk fibroin nanoparticles and the propranolol-encapsulated magnetic silk fibroin nanoparticles; (**b**) hysteresis loop of the magnetic silk fibroin nanoparticles; and (**c**) hysteresis loop of the propranolol-encapsulated magnetic silk fibroin nanoparticles.

FTIR results of silk fibroin-bovine serum albumin nanoparticles and propranolol-encapsulated silk fibroin-bovine serum albumin nanoparticles are presented in Figure 7a,b, respectively, and the characteristic peaks are presented in Table 7. The bands at 1645 cm^{-1} and 1491 cm^{-1} are due to the BSA modification and can be attributed to amide I (C=O stretching) and amide II (C–N stretching and N–H bending) vibrations of BSA, respectively (Table 7). The bands of BSA at 1392 cm^{-1} (–CH_2 bending) and ~1260 cm^{-1} (amide III, C–N stretching, and N–H bending) may be overlapped by the bands attributed to silk fibroin [7]. A slight shift in the spectra for the propranolol-loaded nanoparticles could be due to drug binding.

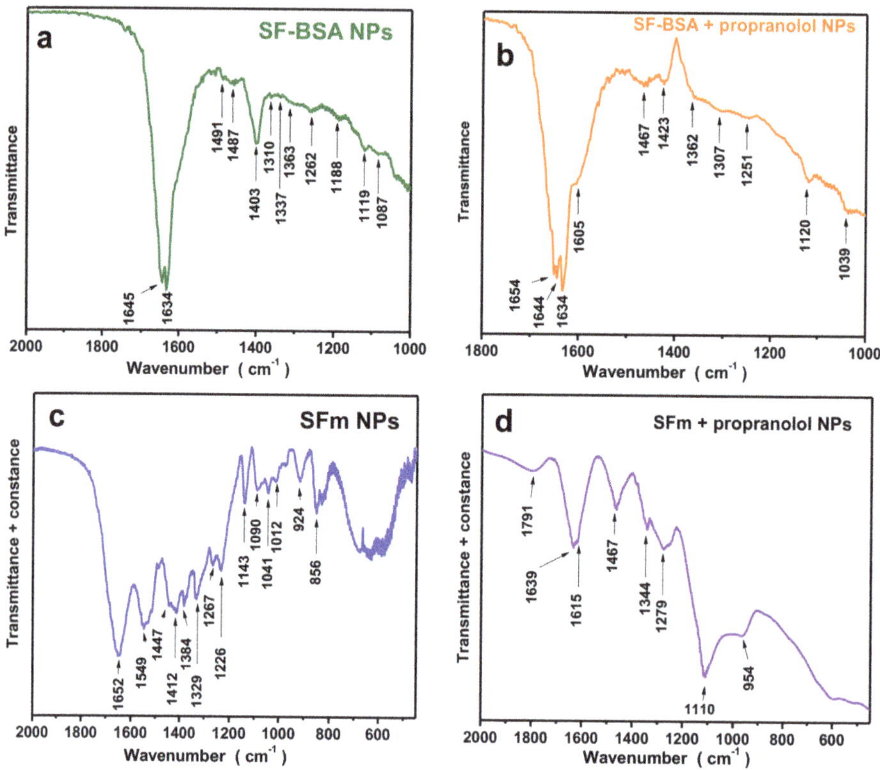

Figure 7. FTIR spectra of: (**a**) silk fibroin-bovine serum albumin nanoparticles; (**b**) propranolol-encapsulated silk fibroin-bovine serum albumin nanoparticles; (**c**) magnetic silk nanoparticles; and (**d**) propranolol-encapsulated magnetic silk fibroin nanoparticles.

In Figure 7c,d, the FTIR results of fibroin magnetic nanoparticles (SFm) and propranolol-encapsulated fibroin magnetic nanoparticles (SFm + propranolol) are presented, while the characteristic peaks are also presented in detail in Table 7. The spectra presented characteristic bands at 471 and 588 cm^{-1} and 506 and 595 cm^{-1}, respectively, attributed to Fe–O bonds [4], which confirmed the successful preparation of magnetic silk fibroin nanoparticles (Figure 7c). From the spectra, the existence of both β-sheets and a-helices for magnetic fibroin nanoparticles could be observed, while for the propranolol-encapsulated silk fibroin magnetic nanoparticles β-sheets could be also observed.

Table 7. Assignment of the main bands present in SF-BSA NPs, SF-BSA + propranolol NPs, and SFm and SFm + propranolol NPs.

cm^{-1}	SF-BSA NPs	cm^{-1}	SF-BSA + Propranolol NPs	cm^{-1}	SFm-NPs	cm^{-1}	SFm + Propranolol NPs
		1652	Amide I random coil	1652	Amide I random coil	1631	Amide I β-sheets
1645	BSA amide I C=O stretching	1644	BSA amide I C=O stretching	1549	Amide II random coil		
1634	Amide I β-sheets	1634	Amide I β-sheets			1466	CH_3 bending
1491	BSA amide II C–N stretching N–H bending			1412	CH_3 bending		
1487	CH_3 bending	1467	CH_3 bending	1384	CH_2 bending $(AG)_n$		
1403	CH_3 bending			1264	Amide III a-helices	1274	Amide III β-sheets
1363	CH_2 bending $(AG)_n$	1362	CH_2 bending $(AG)_n$	1226	Amide III β-sheets		
1262	Amide III a-helices	1261	Amide III a-helices				
1120	CH_2OH polyphenols	1120	CH_2OH polyphenols	1096	C–O stretching C–C stretching polyphenols or tyrosine		
1087	C–O stretching C–C stretching			1041	C–O stretching N–C stretchingSer		
1041	C–O stretching N–C stretching, Ser	1039	C–O stretching N–C stretching, Ser			952	$-CH_2$ $(A)_n$ β-sheets
		952	$-CH_2$ $(A)_n$ β-sheets	924	$-CH_2$ $(A)_n$ β-sheets		
924	$-CH_2$ $(A)_n$ β-sheets			595	Fe–O (magnetite)	595	Fe–O (magnetite)

$(A)_n$, polyalanine; $(AG)_n$, polyalanine glycine [28].

3.5.2. In Vitro Drug Release

In Figure 8, the in vitro drug release profiles for the BSA- and magnetic-nanoparticles-modified silk fibroin nanoparticles are presented compared to the initial silk fibroin nanoparticles. Due to their higher encapsulation efficiency, the propranolol-encapsulated SF-BSA nanoparticles showed a higher drug release rate as compared to the propranolol-encapsulated silk fibroin nanoparticles [13]. In contrast, propranolol-encapsulated magnetic silk fibroin nanoparticles showed less release, which may be attributed to the strong electrostatic interactions of magnetite with the model drug. Another reason for the lower release could be the higher content of β-sheets in the propranolol-loaded silk fibroin magnetic nanoparticles causing a slow degradation of the silk fibroin and finally reducing the drug release rate [26].

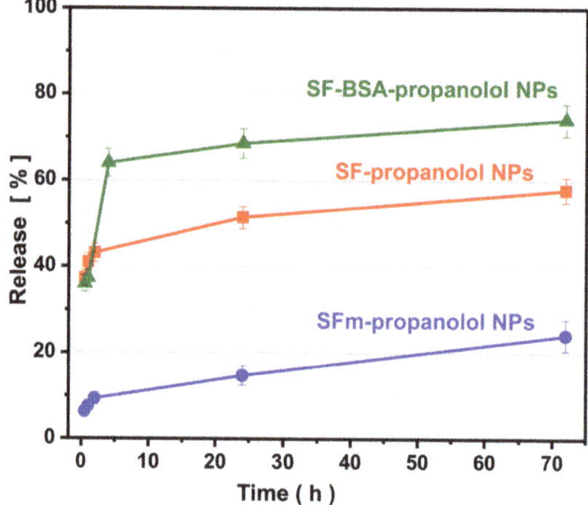

Figure 8. Drug release rate of fibroin nanoparticles, magnetic fibroin nanoparticles, and fibroin–bovine serum albumin nanoparticles encapsulated with propranolol.

4. Conclusions

Silk fibroin nanoparticles were prepared by an easy and safe-to-manipulate method that is amenable to a wide range of drugs and thus useful for silk-based drug delivery systems. Silk fibroin nanoparticles facilitated the entrapment of model drugs (curcumin, pramipexole, and propranolol) with different molecular weights and hydrophobicities and made drug release controllable. Silk fibroin particles loaded with propranolol exhibited a higher release percentage than those loaded with curcumin or pramipexole. Moreover, modified silk fibroin nanoparticles with magnetic nanoparticles as well as with bovine serum albumin were examined for the enhancement of the propranolol encapsulation and release from these nanoparticles. Magnetic silk fibroin nanoparticles increased the propranolol encapsulation efficiency. The silk fibroin-bovine serum albumin particles increased the encapsulation efficiency as well as the release rate of propranolol and were found to be a promising drug carrier.

Author Contributions: Olga Gianak and Eleni Deliyanni conceived and designed the experiments; Olga Gianak performed the experiments; Eleni Pavlidou performed the SEM measurements and analyzed the relative data; Charalampos Sarafidis performed the magnetic experiments and analyzed the relative data; Vassilis Karageorgiou performed the z-dynamic and dynamic light scattering measurements, analyzed the relative data, and contributed to the manuscript's writing; Eleni Deliyanni and Olga Gianak wrote the paper.

Conflicts of Interest: The authors declare no conflict of interest.

References

1. Tudora, M.R.; Zaharia, C.; Stancu, I.C.; Vasile, E.; Trusca, R.; Cincu, C. Natural silk fibroin micro-and nanoparticles with potential uses in drug delivery systems. *UPB Sci. Bull.* **2013**, *75*, 1454–2331.
2. Mottoghitalab, F.; Farokhi, M.; Shokrgozar, M.; Atyabi, M.A.F.; Hosseinkhani, H. Silk fibroin nanoparticle as a novel drug delivery system. *J. Controll. Release* **2015**, *206*, 161–176. [CrossRef] [PubMed]
3. Rahimnejad, M.; Mokhtarian, N.; Ghasemi, M. Production of protein nanoparticles for food and drug delivery system. *Afr. J. Biotechnol.* **2009**, *8*, 4738–4743.
4. Wang, X.; Yucel, T.; Lu, Q.; Hu, X.; Kaplan, D.L. Silk nanospheres and microspheres from silk/pva blend films for drug delivery. *Biomaterials* **2010**, *31*, 1025–1035. [CrossRef] [PubMed]
5. Numata, K.; Kaplan, D.L. Silk-based delivery systems of bioactive molecules. *Adv. Drug Deliv. Rev.* **2010**, *62*, 1497–1508. [CrossRef] [PubMed]

6. Sieb, F.P. Silk nanoparticles—An emerging anticancer nanomedicine. *AIMS Bioeng.* **2017**, *4*, 239–258. [CrossRef]
7. Subia, B.; Kundu, S.C. Drug loading and release on tumor cells using silk fibroin-albumin nanoparticles as carriers. *Nanotechnology* **2013**, *24*, 1–11. [CrossRef] [PubMed]
8. Karimi, M.; Bahrami, S.; Ravari, S.B.; Zangabad, P.S.; Mirshekari, H.; Bozorgomid, M.; Shahreza, S.; Sori, M.; Hamblin, M.R. Albumin nanostructures as advanced drug delivery systems. *Expert Opin. Drug Deliv.* **2016**, *13*, 1609–1623. [CrossRef] [PubMed]
9. Zhao, L.; Zhou, Y.; Gao, Y.; Ma, S.; Zhang, C.; Li, J.; Wang, D.; Li, X.; Li, C.; Liu, Y.; et al. Bovine serum albumin nanoparticles for delivery of tacrolimus to reduce its kidney uptake and functional nephrotoxicity. *Int. J. Pharm.* **2015**, *10*, 180–187. [CrossRef] [PubMed]
10. Tian, Y.; Jiang, X.; Chen, X.; Shao, Z.; Yang, W. Doxorubicin-Loaded Magnetic Silk Fibroin Nanoparticles for Targeted Therapy of Multidrug-Resistant Cancer. *Adv. Mater* **2014**, *26*, 7393–7398. [CrossRef] [PubMed]
11. Prijic, S.; Sersa, G. Magnetic nanoparticles as targeted delivery systems in oncology. *Radiol. Oncol.* **2011**, *45*, 1–16. [CrossRef] [PubMed]
12. Li, H.; Tian, J.; Wu, A.; Wang, J.; Ge, C.; Sun, Z. Self-assembly silk fibroin nanoparticles loaded with binary drug in the treatment of breast carcinoma. *Int. J. Nanomed.* **2016**, *11*, 4370–4380.
13. Gupta, V.; Aseh, A.; Rios, C.N.; Aggarwal, B.B.; Mathur, A.B. Fabrication and characterization of silk fibroin-derived curcumin nanoparticles for cancer therapy. *Int. J. Nanomed.* **2009**, *4*, 115–122. [CrossRef]
14. Kassas, R.A.; Wen, J.; Chen, A.E.M.; Kim, A.M.J.; Liu, S.S.M.; Yu, J. Transdermal delivery of propranolol hydrochloride through chitosan nanoparticles dispersed in mucoadhesive gel. *Carbohydr. Polym.* **2016**, *153*, 176–186. [CrossRef] [PubMed]
15. Leetjens, A.F.G.; Koester, J.; Fruh, B.; Shephard, D.T.S.; Barone, P.; Houben, J.J.G. The Effect of Pramipexole on Mood and Motivational Symptoms in Parkinson's Disease: A Meta-analysis of Placebo-Controlled Studies. *Clin. Ther.* **2009**, *31*, 89–98. [CrossRef] [PubMed]
16. Yu, Z.; Yu, M.; Zhang, Z.; Hong, G.; Xiong, Q. Bovine serum albumin nanoparticles as controlled release carrier for local drug delivery to the inner ear. *Nanoscale Res. Lett.* **2014**, *9*, 2–7. [CrossRef] [PubMed]
17. Rockwwod, D.N.; Preda, R.C.; Yucel, T.; Wang, X.; Lovett, M.L.; Kaplan, D.L. Materials fabrication from *Bombyx mori* silk fibroin. *Nat. Protoc.* **2011**, *6*, 1612–1631. [CrossRef] [PubMed]
18. Shi, P.; Goh, J.C.H. Release and cellular acceptance of multiple drugs loaded silk fibroin particles. *Int. J. Pharm.* **2011**, *420*, 282–289. [CrossRef] [PubMed]
19. Saroyan, H.; Giannakoudakis, D.; Sarafidis, C.; Lazaridis, N.; Deliyanni, E. Effective impregnation for the preparation of magnetic mesoporous carbon: Application to dye adsorption. *J. Chem. Technol. Biotechnol.* **2017**, *92*, 1899–1911. [CrossRef]
20. Leucuta, S.E.; Bodea, A. Optimization of propranolol hydrochloride sustained release pellets using a factorial design. *Int. J. Pharm.* **1997**, *154*, 49–57.
21. Li, M.; Lu, S.; Wu, Z.; Tan, K.; Minoura, N.; Kuga, S. Structure and properties of silk fibroin-poly-vinyl alcohol gel. *Int. J. Biol. Macromol.* **2002**, *30*, 89–94. [CrossRef]
22. Zhang, H.; Li, L.L.; Dai, F.Y.; Zhang, H.H.; Ni, B.; Zhou, W.; Yang, X.; Wu, Y.Z. Preparation and characterization of silk fibroin as a biomaterial with potential for drug delivery. *J. Transl. Med.* **2012**, *10*, 1–9. [CrossRef] [PubMed]
23. Zhao, Z.; Li, Y.; Xie, M.B. Silk fibroin-based nanoparticles for drug delivery. *Int. J. Mol. Sci.* **2015**, *16*, 4880–4903. [CrossRef] [PubMed]
24. Kamalha, E.; Zheng, Y.; Zeng, Y. Analysis of the secondary crystalline structure of regenerated *Bombyx mori* fibroin. *Res. Rev. Biosci.* **2013**, *7*, 76–83.
25. Huang, X.; Fan, S.; Altayp, A.I.; Zhang, Y.; Shao, H.; Hu, X.; Xie, M.; Xu, Y. Tunable Structures and Properties of Electrospun Regenerated Silk Fibroin Mats Annealed in Water Vapor at Different Times and Temperatures. *J. Nanomater.* **2014**, 1–7. [CrossRef]
26. Reddy, M.K.D.; Sriniva, N.; Shanker, P. Formulation development and in vitro evaluation of pramipexole dihydrochloride monohydrate extended release matrix tablets. *Int. J. Pharm. Sci. Res.* **2014**, *2*, 2119–2133.
27. Xie, M.B.; Li, Y.; Zhao, Z.; Chen, A.Z.; Li, J.S.; Hu, J.Y.; Li, G.; Li, Z. Solubility enhancement of curcumin via supercritical CO_2 based silk fibroin carrier. *J. Supercrit. Fluids* **2015**, *103*, 1–9. [CrossRef]
28. Kush, P.; Thakur, V.; Kumar, P. Formulation and in vitro evaluation of propranolol hydrochloride loaded polycaprolactone microspheres. *Int. J. Pharm. Sci. Rev. Res.* **2013**, *20*, 282–290.

29. Audet, M.B.; Volrath, F.; Holland, C. Idnetification and classification of silks using infrared spectroscopy. *J. Exp. Biol.* **2015**, *218*, 3138–3149. [CrossRef] [PubMed]
30. Yu, S.; Yang, W.; Chen, S.; Chen, M.; Liu, Y.; Shao, Z.; Chen, X. Floxuridine-loaded silk fibroin nanospheres. *RSC Adv.* **2014**, *4*, 18171–18177. [CrossRef]

Technical Note

Method Development of Phosphorus and Boron Determination in Fertilizers by ICP-AES

Emanouela Viso and George Zachariadis *

Laboratory of Analytical Chemistry, Department of Chemistry, Aristotle University, 54124 Thessaloniki, Greece; emmyviso@gmail.com

* Correspodence: zacharia@chem.auth.gr; Tel.: +30-231-099-7707

Received: 14 May 2018; Accepted: 3 July 2018; Published: 9 July 2018

Abstract: Simultaneous determination of phosphorus and boron in fertilizers was performed by Inductively Coupled Plasma Atomic Emission Spectroscopy (ICP-AES). Three different samples were analyzed, of which two were inorganic and one was of organic composition. Analysis of the samples was performed after heated acidic digestion to completely dissolve them, using two different acid mixtures. A solution of HCl + HNO_3 was used to digest the inorganic fertilizers, and a solution of H_2SO_4 + HNO_3 for the organic fertilizer. The spectral emission lines used were 213.617 nm and 214.917 nm for P and 249.772 nm, 249.677 nm and 208.957 nm for B. The detection and quantification limits for P were between 10–20 mg/kg and 40–80 mg/kg respectively, while for B they ranged between 10–30 mg/kg and 40–100 mg/kg respectively. The repeatability of the technique was found to be within the range 0.9–17.0% for P and 1.7–23.4% for B, expressed as relative standard deviation (RSD). The concentrations found by the proposed method are in good agreement with those reported on their package labels.

Keywords: phosphorus; boron; inductively coupled plasma; atomic emission spectrometry; fertilizers; acid dissolution; wet digestion

---

1. Introduction

Fertilizers contain many elements which act as nutrients for plants, but also elements which may be toxic, and which are therefore responsible for contamination of soils. This results in the accumulation of chemicals in food [1,2]. As the development of agricultural productivity is directly related to the use of fertilizers, it is necessary to analyze them with sensitive, multi-elemental technical analyzers to monitor their quality [3]. Common fertilizers are either of inorganic or organic composition, natural or synthetic. Chemical analysis and quality control of fertilizers result in improved agricultural production [4], and methods have been developed. In this context, the determination of minerals in fertilizers becomes necessary, and for this purpose, several techniques have been reported in the literature [1,2,5]. Some techniques for the determination of heavy metals are based on Graphite Furnace atomic absorption spectroscopy (GF-AAS), Continuous Source Flame Absorption Spectroscopy (HR-CS-FAAS) [1,6–9], or Cold Vapor Atomic Absorption Spectroscopy (CV-AAS) for Hg [5]. In addition, methods have been developed to determine nutrients in fertilizers by LIBS [10,11], or by total X-ray fluorescence (TXRF) techniques [12]. Inductively coupled plasma atomic emission spectroscopy (ICP-AES) allows rapid analysis and simultaneous determination of primary and secondary nutrients as trace elements in fertilizers [2,13,14]. ICP-AES is a well-known analytical technique suitable for simultaneous determination of almost 70 chemical elements, as it provides high sensitivity and satisfactory detectability, has a wide range of applications, and does not suffer from important spectral interferences [15–17], except in case of solutions with high TDS. The great advantage of simultaneous determination of various elements in fertilizers, like P and K, has already been highlighted by other researchers [18]; this becomes more attractive considering the capability of measuring also semi-metals, like boron [19], or non-metals, like phosphorus and sulfur. However, in many cases for total element fraction, it requires complete digestion of solid samples before quantitative analysis, which means a sample treatment step, which can potentially affect accuracy and overall analytical performance of the method. In the present study, a method for the simultaneous determination of total phosphorus and boron by ICP-AES in fertilizers was developed, after digestion of the samples with two different acidic solutions. The examined acid mixtures were selected according to the nature of the fertilizer samples. For determination of the total fraction of an element, inorganic fertilizers usually do not require strong oxidative media, while organic fertilizers may need strong oxidative acids and heating or incineration followed by acidic dissolution for complete decomposition or other pretreatment techniques [19,20].

2. Experimental

2.1. Instrumentation

An inductively coupled plasma atomic emission spectrophotometer, model OPTIMA 3100 XL (Perkin Elmer, San Francisco, CA, USA), was employed; the operating conditions are listed in Table 1. The instrument was equipped with a liquid sample introduction system, including a peristaltic pump to aspirate solutions into the nebulizer at variable flow rates. Tygon-type pump tubes were used for sample delivery, and the nebulization system consisted of a cross-flow nebulizer and a Scott double-pass spraying chamber. A quartz torch is mounted horizontally on the same axis to the spectrophotometer window (axial viewing of emission). The torch inner injector tube was composed of alumina, which is sufficiently resistant to highly acidic solutions. The radio frequency generator (RF) that maintains the plasma is 40.68 MHz; this was adjusted to an incident power of 1350 watts for this method. The spectrophotometer was equipped with a polychromator in which an echelle-type diffraction grating was installed, and the detector was a segmented charged-coupled device. The ICP-AES instrument was supplied by analytical grade argon as a plasma gas. Three phosphorous and three boron atomic spectral lines were examined, as given in Table 2. The phosphorous line at 178.221 nm was finally rejected for reasons described below.

Table 1. Operating conditions and settings of the ICP-AES.

Parameter	Value
RF generator	40.68 MHz, free-running
RF incident power	1350 W
Torch type	Fassel-type
Injector, id	Alumina 2.0 mm
Viewing mode	Axial
Auxiliary argon flow rate	0.50 L min^{-1}
Nebulizer argon flow rate	0.80 L min^{-1}
Plasma gas flow rate	15 L min^{-1}
Spray chamber type	Scott
Sample uptake flow rate	2 mL min^{-1}
Detector	Segment-array charge-coupled (SCD)

Table 2. Spectral lines employed for ICP-AES measurements.

Element	Spectral Lines (nm)		
P	178.221	213.616	214.914
B	208.977	249.677	249.772

2.2. Reagents and Solutions

For the preparation of the working standard solutions, KH_2PO_4 (99.5%) and H_3BO_3 (99.5%) were obtained from Merck (Darmstadt, Germany). For acid digestion of fertilizer samples, concentrated HNO_3 (65%), HCl (37%), and H_2SO_4 (95–97%) of analytical grade were obtained from Merck (Darmstadt, Germany). All dilutions and solutions were carried out using ultrapure water of Milli-Q quality (18.2 MΩ, Millipore, Bedford, MA, USA).

2.3. Fertilizer Samples

Inorganic and organic commercially available fertilizers from local companies in Greece were analyzed in the present study to develop a fast analytical method. The inorganic samples were crystalline, water-soluble, general-purpose fertilizers, while the organic one was a composite fertilizer for vegetables. Their indicative nutritional content, as given on their packaging, is provided in Table 3. Phosphorus and boron contents of the samples were calculated, and are presented in Table 4.

Table 3. Nutritional content of the three fertilizer samples.

Samples	Chemical Compositions										
	N % *w/w*	P (P_2O_5) % *w/w*	K (K_2O) % *w/w*	Mg (MgO) % *w/w*	Fe mg/kg	Cu mg/kg	Mo mg/kg	B mg/kg	Mn mg/kg	Zn mg/kg	Co mg/kg
1st sample (*inorganic*)	20	20	20	0.09	400	150	1	150	400	150	5
2nd sample (*inorganic*)	5	10	42	4	500	100	100	100	100	500	-
3rd sample * (*organic*)	4	30	-								

* In 3rd sample (organic) organic matter is 15% *w/w*.

Table 4. Concentration of P and B in the samples.

Element	Mass Concentration g/kg		
	1st Sample (*Inorganic*)	2nd Sample (*Inorganic*)	3rd Sample (*Organic*)
P	87	43	131
B	0.15	0.10	-

2.4. Preparation of Working Standard Solutions

A mixed standard aqueous solution containing 50 mg/L each of phosphorus and boron was prepared from standard solutions of 100 mg/L P (KH_2PO_4, 439.4 mg/L) and 100 mg/L B (H_3BO_3, 571.97 mg/L). Finally, five working standard solutions containing 0.00, 1.00, 2.50, 10.0, 25.0 mg/L of phosphorus and boron, respectively, were prepared by proper dilutions of the above stock standard solution (50 mg/L).

2.5. Acid Digestion of Samples

For the selection of the appropriate acid mixtures, dissolution tests were performed by acid digestion of all samples. The most efficient mixtures were selected for the liquid digestion process based on the solubilization effect. Thus, for the 1st and the 2nd sample (inorganic), the acid digestion procedure was performed with the addition of 5 mL of 37% HCl + 1.5 mL of 65% HNO_3, and then heating for 2 min on a hot plate in high fume hood (ca 80 fpm face velocity) for nitrogen oxides fumes. For the 3rd sample (organic), the acid digestion procedure was performed by adding 5 mL H_2SO_4 95–97% + 1 mL HNO_3 65%, followed by heating for 3 min on a hot plate in a fume hood (ca 80 fpm). For the acid digestion procedure, accurately weighed amounts of the samples were placed in 100 mL open conical flasks and heated to 130 °C. When digestion was complete, the mixture was allowed to cool to room temperature. The mixtures were then transferred to 100 mL volumetric flasks and diluted with Milli-Q water. The obtained diluted solution was further diluted at a ratio of 1:10 and 1:100 successively. The acid digestion process was repeated three times for each sample. Preparation of acidic blank mixtures was performed three times.

Standard addition samples were also prepared by the addition of standard solutions of P and B to 0.500 g of the first sample. This process was carried out to study the recovery of the analytical method. To this sample, 228 g/kg P (435 mg/L) of KH_2PO_4 and 174 g/kg B (0.75 mg/L) of H_3BO_3 were added. The sample was then subjected to acidic digestion with 5 mL HCl + 1.5 mL HNO_3 and warmed up. The sample after acid digestion was diluted to 100 mL, followed by successive dilutions of 1:10 and 1:100. The whole procedure was performed three times.

3. Results and Discussion

3.1. Regression Analysis

The settings for the spectral lines and the operating conditions of ICP-AES were defined in the computer software. For quantitative analyses of the samples, the required calibration curves were constructed using the series of mixed working standard solutions.

The results of the regression analysis for each element at each tested wavelength are given in Table 5. The emission spectra of P and B obtained from the mixed working standard solutions are given in Figures 1 and 2 respectively. As shown in Table 5, at the examined spectral lines, both P and B showed good correlation coefficients. The correlation coefficient R is greater than 0.999 for all elements, except at the third spectral line of P (178.221 nm), where the correlation was much lower because of baseline instability. Apparently, no sensitivity is observed also, and for this reason, this spectral line P (178.221 nm) was excluded from further measurements.

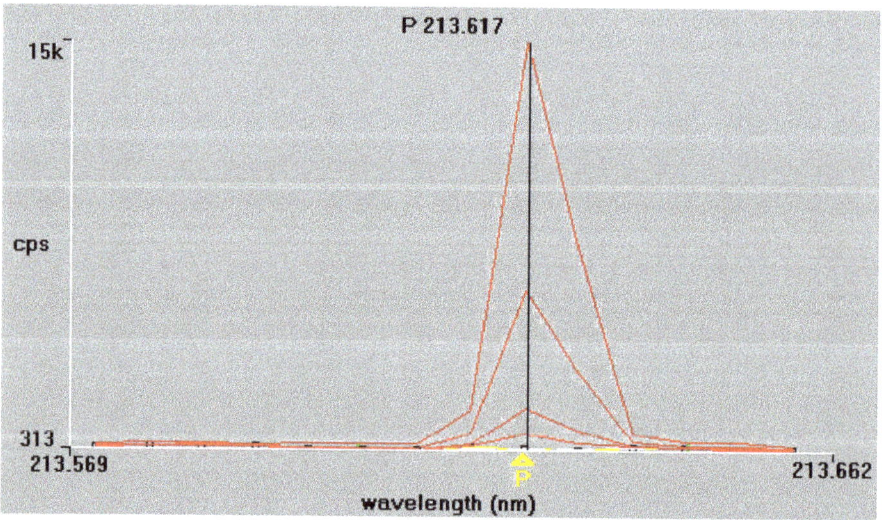

Figure 1. *Cont.*

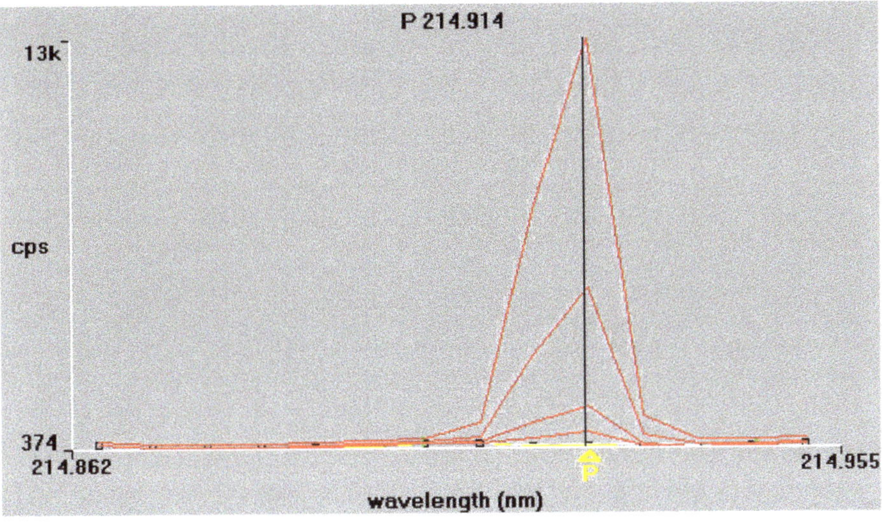

Figure 1. Phosphorus emission spectra at 213.617 nm and 214.914 nm, respectively. Superimposed traces refer to working standard solutions of increasing concentration, as described in Section 2.4.

Table 5. Results of regression analysis using the mixed standard aqueous solutions of boron and phosphorus.

Element	Spectral Line (nm)	Slope [cps/(mg/L)]	Standard Error	Correlation Coefficient R
P	213.616	621 ± 19	61	0.9999
	214.914	777 ± 38	123	0.9999
	178.221	−0.5 ± 2	5	0.7171
B	249.772	53,417 ±1644	5390	0.9999
	249.677	19,771 ± 496	1627	0.9999
	208.977	5274 ± 161	530	0.9999

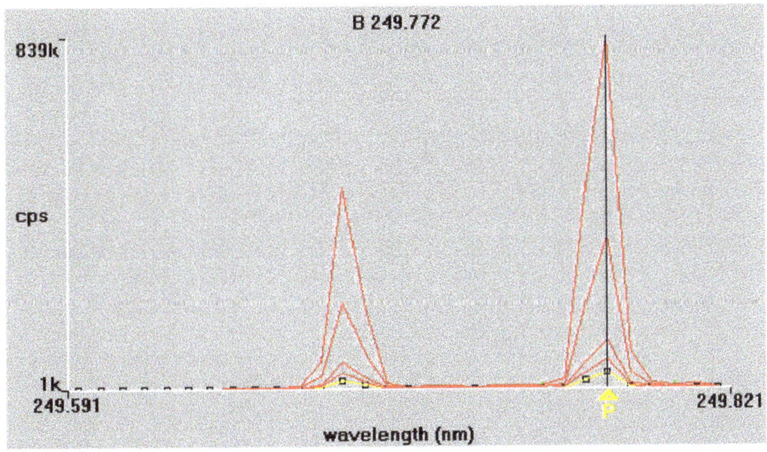

Figure 2. *Cont.*

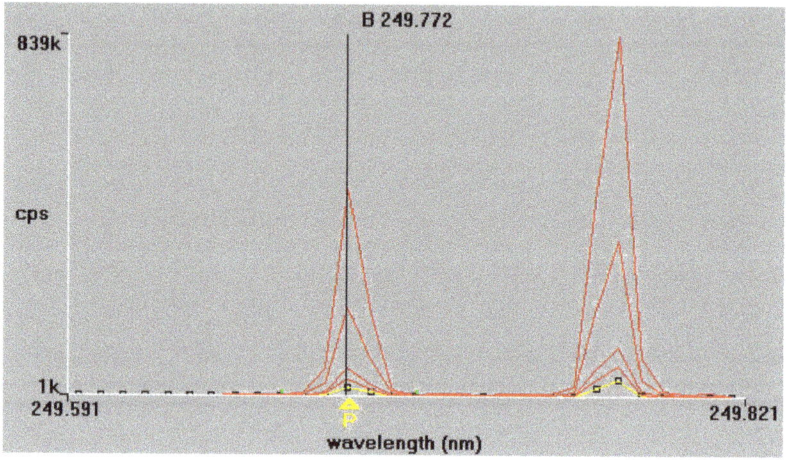

Figure 2. Boron emission spectra at 249.772 nm and 249.677 nm respectively. Superimposed traces refer to standard solutions of increasing concentration, as described in Section 2.4.

3.2. Repeatability, Detectability and Recovery of Method

Acceptable repeatability of a method is expressed by low values of the relative standard deviation, RSD%. Repeatability testing was done by measuring the working standard three times on the same day. The solutions used were aqueous mixed solutions of P and B on all spectral lines. The concentrations used to control the repeatability were 1.00 mg/L, 2.50 mg/L, 10.0 mg/L, and 25.0 mg/L. The results are listed in Table 6. Based on these results, good repeatability is observed at concentrations of 10.0 mg/L and 25.0 mg/L, but less good at 1.00 mg/L concentration. The repeatability for P in the examined atomic lines ranged between 0.9–17%, and for B between 1.7–23.4%, respectively.

Limit of detection (LOD) and limit of quantification (LOQ) are defined as the minimum concentration or quantity of the analyte that can be detected or quantified respectively with reasonable certainty, according to IUPAC recommendations [21]. The detection and quantification limits of the developed method were calculated using the 3s criterion (i.e., at a 99.6% confidence level). The calculated detection limits for the examined spectral lines of phosphorus were found to be between 10–20 mg/kg, and those for B between 10–30 mg/kg. This variation is due to variability of the baseline signal values in different spectral lines and with the different acid mixtures. The corresponding limits of quantification were found to be between 40–80 mg/kg for P, and between 40–100 mg/ kg for B. The results are presented in Tables 7 and 8. In these tables, the results are given as typical instrumental LODs and LOQs (expressed in mg/L), and also as method equivalent LODs and LOQs (expressed in mg/kg), considering a typical sample mass of 0.500 g for the analysis and initial dilution to 100 mL, as described in paragraph 2.5.

Because no reference material was available, the recovery for P and B was calculated after adding an amount of the standard solution to the first sample of fertilizer. Table 9 gives the calculated recoveries for P and B in each spectral line, which are satisfactory regarding the spectral lines 213.616 nm for P and 249.772 nm for B, correspondingly. The added amount of the standard was 228 g/kg (or 435 mg/L) for P (as KH_2PO_4), and 176 g/kg (or 0.75 mg/L) for B (as H_3BO_3).

Table 6. Method repeatability.

Element	Spectral Line (nm)	RSD %			
		c = 1.00 mg/L	c = 2.50 mg/L	c = 10.0 mg/L	c = 25.0 mg/L
P	213.616	14.8	4.9	0.9	2.3
	214.914	17.0	6.2	1.3	2.9
B	249.772	23.4	7.9	1.7	2.0
	249.677	19.4	7.3	1.9	1.8
	208.977	18.6	6.5	1.9	2.2

Table 7. Detection and quantification limits using a constitutive blank of 3 mL HCl + 1 mL HNO_3. LODs and LOQs were calculated as described in Section 3.2.

Element	Spectral Line (nm)	LOD (mg/L)	LOD (mg/kg)	LOQ (mg/L)	LOQ (mg/kg)
P	213.616	0.07	14	0.30	60
	214.914	0.06	12	0.20	40
B	249.772	0.10	20	0.50	100
	249.677	0.10	20	0.40	80
	208.977	0.15	30	0.50	100

Table 8. Detection and quantification limits using a constitutive blank of 5 mL H_2SO_4 + 1 mL HNO_3. LODs and LOQs were calculated as described in Section 3.2.

Element	Spectral Line (nm)	LOD (mg/L)	LOD (mg/kg)	LOQ (mg/L)	LOQ (mg/kg)
P	213.616	0.10	20	0.40	80
	214.914	0.05	10	0.20	40
B	249.772	0.10	20	0.40	80
	249.677	0.06	12	0.20	40
	208.977	0.05	10	0.20	40

Table 9. Recovery results.

Element	Spectral Line (nm)	Recovery R%
		Adds: 228 g/kg P + 174 g/kg B (435 mg/L P + 0.75 mg/L B)
P	213.616	97
	214.914	<88
B	249.772	98
	249.677	>112
	208.977	>112

4. Application of the Method to Fertilizers Samples

The results of the analysis of elements P and B in three samples of inorganic and organic fertilizer samples are presented in Table 10, while the corresponding label compositions are given in Tables 3 and 4. Acid digestion was performed three times for each sample under the same conditions and with the same amounts of samples and solvents. Then, each of the three sets of samples was diluted three times, i.e., initial dilution to 100 mL, which was used for boron determination, and subsequent dilutions 1:10 and 1:100, which were used for phosphorus determination.

Regarding the 1st inorganic fertilizer, the indicated content for P in the form P_2O_5 is 20% *w*/*w*. Thus, the concentration expressed in elemental P is 87 g/kg. The concentration using either of the two spectral lines after the analysis is in good accordance with that on the label. On the other hand, the concentration of B as given on the label is 0.015% *w/w*, which means 0.15 g/kg. The concentration as found after analysis i of the second and third spectral lines agrees with the indicated value.

Regarding the 2nd inorganic fertilizer, the indicated content for P given in the form of P_2O_5 is 10% *w/w*, so the concentration of elemental P is 43 g/kg. This concentration is also in agreement with the concentration resulting from sample analysis using the first spectral line of P. The B content given on the label is 100 ppm, which means 0.10 g/kg. The concentrations found from the first spectral line of boron are consistent with the reported content.

Finally, the reported content of the 3rd organic fertilizer sample is 30% *w/w* P_2O_5; therefore, the elemental P concentration is expected to be 131 g/kg, which is in accordance with the concentrations resulting from ICP-AES analysis. For B, its content is not indicated on the package. The concentration from the ICP-AES analysis was found to be at 0.12 ± 0.03 g/kg using the first emission line.

Table 10. Concentration of P and B in fertilizer samples found by ICP-AES analysis after digestion.

Element	Spectral Line (nm)	Concentration g/Kg		
		1st Inorganic Fertilizer	2nd Inorganic Fertilizer	3rd Organic Fertilizer
P	213.616	100 ± 40	34 ± 7	129 ± 9
	214.914	65 ± 11	23 ± 5	101 ± 18
B	249.772	0.10 ± 0.01	0.13 ± 0.03	0.12 ± 0.03
	249.677	0.17 ± 0.01	0.21 ± 0.04	0.27 ± 0.02
	208.977	0.13 ± 0.01	0.16 ± 0.04	0.15 ± 0.02

5. Conclusions

In this study, the ICP-AES technique was applied to the development of a multi-elemental fertilizer analytical method. The sample solution after heated acid wet digestion was introduced into the inductively coupled plasma atomizer. From the conducted research, the following conclusions regarding the performance of the developed method were drawn. The spectral line of P (178.221 nm) showed low correlation coefficient and poor sensitivity. The spectral lines P (213.16 nm) and P (214.914 nm) showed very good correlation, R > 0.999, with spectral line 214.914 nm being preferable in terms of sensitivity. The three spectral lines of B showed similar correlation coefficients, with a spectral line at 249.677 nm being preferable to the others in terms of sensitivity. The repeatability at high concentration levels of the working curve is better compared to the lower concentrations, and the calculated recoveries were satisfactory. Wet acid digestion of samples was completed in short pretreatment time using mixtures of HCl, HNO_3, and H_2SO_4 acids. The developed method was applied to three different fertilizers of inorganic and organic natures. The P and B results are in accordance with those reported on the product packages. According to the results, it has been concluded that the developed method is a reliable, simple, and fast method for the simultaneous determination of P and B in fertilizers. The method can be applied to rapid examination during fertilizers production, as well as in the analysis of commercial products.

Author Contributions: Conceptualization and Investigation G.Z.; Methodology E.V.; Writing-Original Draft Preparation, E.V.; Writing-Review & Editing, G.Z.; Supervision, G.Z.

Funding: This research received no external funding.

Conflicts of Interest: The authors declare no conflict of interest.

References

1. Borges, A.R.; Becker, E.M.; Lequeux, C.; Vale, M.G.R.; Ferreira, S.L.C.; Welz, B. Method development for the determination of cadmium in fertilizers samples using high-resolution continuum source graphite furnace atomic absorption spectrometry and slurry sampling. *Spectrochim. Acta B* **2011**, *66*, 529–535. [CrossRef]

2. Kane, P.F.; Hall, W.L., Jr. Determination of arsenic, cadmium, cobalt, chromium, lead, molybdenum, nickel, and selenium in fertilizers by microwave digestion and inductively coupled plasma-optical emission spectrometry detection: Collaborative study. *J. AOAC Int.* **2006**, *89*, 1447–1466. [PubMed]
3. Soil Science Society of America. Glossary of Soil Science Terms. 2008. Available online: https://www.soils.org/publications/soils-glossary (accessed on 20 April 2018).
4. Das, S.; Adhya, T.K. Effect of combine of inorganic manure and inorganic fertilizer on methane and nitrous oxide emissions from a tropical flooded soil planted to rice. *Geoderma* **2014**, *213*, 185–192. [CrossRef]
5. Zhao, X.; Wang, D. Mercury in some chemical fertilizers and the effect of calcium superphosphate on mercury uptake by corn seedlings (*Zea mays* L.). *J. Environ. Sci.* **2010**, *22*, 1184–1188. [CrossRef]
6. De Morais, C.P.; Barros, A.I.; Santos, J.D.; Ribeiro, C.A.; Crespi, M.S.; Sanesi, G.S.; Neto, J.A.G.; Ferreira, E.C. Calcium determination in biochar-based fertilizers by laser-induced. *Microchem. J.* **2007**, *134*, 370–373. [CrossRef]
7. De Oliveira Souza, S.; Froncois, L.L.; Borges, A.R.; Vale, M.G.R.; Araujo, R.G.O. Determination of copper and mercury in phosphate fertilizers employing direct solid sampling analysis and high resolution continuum source graphite furnace atomic absorption spectrometry. *Spectrochim. Acta Part B At. Spectrosc.* **2015**, *114*, 58–64. [CrossRef]
8. Borges, A.R.; Francois, L.L.; Becker, E.M.; Vale, M.G.R.; Welz, B. Method development for the determination of chromium and thallium in fertilizers samples using graphite furnace atomic absorption spectrometry and direct solid samples analysis. *Microchem. J.* **2015**, *119*, 169–175. [CrossRef]
9. Bechlin, M.A.; Fortunato, F.M.; de Silva, R.M.; Ferreira, E.C.; Neto, J.A.G. A simple and fast method for assessment of the nitrogen-phosphorus-potassium rating of fertilizers using high-resolution continuum source atomic and molecular absorption spectrometry. *Spectrochim. Acta Part B At. Spectrosc.* **2014**, *101*, 240–244. [CrossRef]
10. Nunes, L.C.; Gustinelli, C.; de Carvallo, A.; Santos, J.D.; Krug, F.J. Determination of Cd, Cr and Pb in phosphate fertilizers by laser- induced breakdown spectroscopy. *Spectrochim. Acta Part B At. Spectrosc.* **2014**, *97*, 42–48. [CrossRef]
11. Nicolodelli, G.; Senesi, G.S.; Perazzoli, I.L.O.; Marangoni, B.S.; Benites, U.D.M.; Milari, D.M.B.P. Double pulse laser induced breakdown spectroscopy: A potential tool for the analysis of contaminants and macro/micronutrients in organic mineral fertilizers. *Sci. Total Environ.* **2016**, *565*, 1116–1123. [CrossRef] [PubMed]
12. Resendes, L.V.; Nascentes, C.C. A simple method for the multi-elemental analysis of organic fertilizer by slurry sampling and total reflection X-ray fluorescence. *Talanta* **2016**, *147*, 485–492. [CrossRef] [PubMed]
13. Nziguheba, G.; Smolders, E. Inputs of trace elements in agricultural soils via phosphate fertilizers in European countries. *Sci. Total Environ.* **2008**, *390*, 53–57. [CrossRef] [PubMed]
14. Rui, Y.; Hao, J.; Rui, F. Determination of seven plant nutritional element in potassium dihydrogen phosphate fertilizers from northeastern China. *J. Saudi Chem. Soc.* **2012**, *16*, 89–90. [CrossRef]
15. Santos, J.S.; Teixeira, L.S.G.; Araujo, R.G.O.; Fernandes, A.P.; Korn, M.G.A.; Ferreira, S.L.C. Optimization of the operating conditions using factorials design for determination of uranium by inductively plasma optical emission spectrometry. *Microchem. J.* **2011**, *97*, 113–117. [CrossRef]
16. De Oliveira Souza, S.; de Costa, S.L.; Santos, D.M.; Pinto, J.D.S.; Garcia, C.A.B.; Alves, J.D.P.H.; Araujo, R.G.O. Simultaneous determination of macronutrients, micronutrients and trace elements in mineral fertilizers by inductively coupled plasma optical emission spectrometry. *Spectrochim. Acta Part B At. Spectrosc.* **2014**, *96*, 1–7. [CrossRef]
17. Zachariadis, G. Inductively Coupled plasma atomic emission spectometry, A Model. In *Multi-Elemental Technique for Modern Analytical Laboratory*; Nova Science Publishers: New York, NY, USA, 2011; pp. 18–65.
18. Bartos, J.M. Determination of Phosphorus and Potassium in Commercial Inorganic Fertilizers by Inductively Coupled Plasma-Optical Emission Spectrometry: Single-Laboratory Validation. *J. AOAC Int.* **2014**, *97*, 687–699. [CrossRef] [PubMed]
19. Kimura, M. *Testing Methods for Fertilizers*; Food and Agricultural Materials Inspection Center (FAMIC): Saitama, Japan, 2016; pp. 92–111, 259–269.

20. Faithfull, N.T. The analysis of fertilizers, Cp. 6. In *Methods in Agricultural Chemical Analysis: A Practical Handbook*; CABI Publishing: Oxon, UK, 2002; pp. 110–118.
21. International Union of Pure and Applied Chemistry (IUPAC). *Compendium in Chemical Terminology, Version 2014*; Blackwell Scientific Publications: Oxford, UK, 1997. Available online: https://goldbook.iupac.org/html/L/L03540.html (accessed on 23 June 2018).

Article

NSAIDs Determination in Human Serum by GC-MS

Adamantios Krokos [1,2], Elisavet Tsakelidou [2], Eleni Michopoulou [1], Nikolaos Raikos [2], Georgios Theodoridis [1] and Helen Gika [2,*]

1 Laboratory of Analytical Chemistry, Department of Chemistry, Aristotle University of Thessaloniki, Thessaloniki 54124, Greece; akrokosa@chem.auth.gr (A.K.); elena.michopoulou@gmail.com (E.M.); gtheodor@chem.auth.gr (G.T.)

2 Laboratory of Forensic Medicine & Toxicology, Department of Medicine, Aristotle University of Thessaloniki, Thessaloniki 54124, Greece; tsakeliz@hotmail.com (E.T.); raikos@med.auth.gr (N.R.)

* Correspondence: gkikae@auth.gr; Tel.: +30-2310-996-224

Received: 30 April 2018; Accepted: 3 July 2018; Published: 16 July 2018

Abstract: Non-steroidal anti-inflammatory drugs (NSAIDs) are being widely consumed without medical prescription and are often the cause of intoxication, usually in young children. For this, there is a special need in their determination in routine toxicology analysis. As screening methods mainly focus on drugs of abuse (DOA) that are alkaline compounds in their majority, they are not optimized for acidic drugs, such as NSAIDs. Thus, more specific methods are needed for the detection and quantification of this class of drugs. In this study, the efficient extraction of NSAIDs from blood serum and their accurate determination is studied. Optimum pH extraction conditions were studied and thereafter different derivatization procedures for their detection. From the derivatization reagents used, *N*,*O*-Bis(trimethylsilyl)trifluoroacetamide (BSTFA) with 1% Trimethylchlorosilane (TMCS) was found to be the optimum choice for the majority of the examined NSAIDs; pH of 3.7 was selected as the most efficient for the extraction step. Herein the formation of the lactam of diclofenac was also thoroughly investigated. The developed Gas Chromatography-Mass Spectrometry (GC-MS) method had a run time of 15 min with the mass spectrometer operating in Electron Impact (EI) within the mass range of 40 to 500 amu. The method was linear with R^2 above 0.991 and limits of quantitation (LOQ) ranging from 6 to 414 ng/mL. The intra-day accuracy and precision were found between 1.03%–9.79% and 88%–110%, respectively, and the inter-day accuracy and precision were between 1.87%–10.79% and 91%–113%. The optimum protocol was successfully applied to real clinical samples, where intoxication of NSAIDs was suspected.

Keywords: NSAIDs; derivatization; GC-MS; serum

1. Introduction

Non-steroidal anti-inflammatory drugs (NSAIDs) are acidic compounds with anti-inflammatory properties at high concentrations and several other properties at low concentrations (i.e., salicylic acid is used as anticoagulant drug) [1]. These drugs exhibit toxicity in the upper concentration levels or after a long time of intake [2,3]. Some of the toxic side effects are related with gastrointestinal disorders, intestinal ulceration, aplastic anemia, myocardial infarction, cerebrovascular events, inhibition of platelet aggregation, and renal dysfunction [4]. Especially, acetaminophen and nimesulide exhibit significant hepatotoxicity [2,5–7]. In addition, there are several cases of suicide attempts or crime commissions which are related with NSAIDs [8].

There is a number of analytical methods which report NSAIDs' determination in biological fluids, most of them related to pharmacokinetic studies or to the support of animal studies for the estimation of exposure and the investigation of potential risk after consumption [4,9].

There are reports of determinations of these analytes by immunoassays, Gas Chromatography-Flame Ionization Detection (GC-FID), High Performance Liquid Chromatography-UltraViolet/Diode

Array Detection (HPLC-UV/DAD)/fluorimetric detector, spectrofluorimetry, thin layer chromatography-UV/fluorimetric detector, GC-MS, GC-MS/MS, capillary electrophoresis, LC-MS, and LC-MS/MS [10–14]. Immunoassays suffer from low selectivity due to cross-reactions, while all other methods except LC-MS/MS include time-consuming sample pretreatment. There is a good number of published multi-analyte LC-MS/MS methods for measuring NSAIDs and their metabolites that provide sensitive and reliable concentration data from biological matrices [5]. In particular, those methods are the most suitable for low concentrated salicylates originating from nutrition. However, GC-MS still represents an integral tool of a clinical and/or toxicological laboratory due to the fact that mass spectra databases are available aiding in the identification of unknowns, At the same time, latest instruments and materials hold the promise for better chromatographic separations and sensitivity, critically important for complex biological samples. Thus, GC-MS analysis provides advantages clearly important in such applications.

For thedetection of NSAID by GC-MS, various sample pretreatment protocols have been applied. Acidic liquid-liquid extraction is often used, however some researchers have also developed simple SPE protocols instead [3,11]. Furthermore, it has been shown that the selection of a suitable derivatization agent is a key factor for the sensitivity of their detection. Apart from this, there are several issues which are related with the stability of these compounds in the GC-MS conditions.

In this paper, a multi-analyte GC-MS method was developed for the simultaneous quantification of eight NSAIDs, based on a specific pro-analysis procedure. The optimum conditions were selected based on experiments focusing on the sample treatment, to obtain a method able to address the needs of a toxicological analysis for these challenging acidic analytes. The method provides a valuable tool for the determination of eight different NSAIDS by GC-MS, with the pros that the GC-MS technique offers, as stressed above, the additional potential of identification of metabolites, degradation products, or others.

2. Materials and Methods

2.1. Chemicals and Reagents

All solvents were of analytical or LC-MS grade. Acetonitrile (ACN) LC-MS was purchased from Chem Lab NV, (Zedelgem, Belgium). Nordiazepam-d_5 solution 1 mg/mL in methanol reference, material reference standards (>99%) of acetaminophen (APAP), acetyl salicylic acid (ASA), salicylic acid (SA), ibuprofen (IBP), diclofenac (DCF), nimesulide (NI), niflumic acid (NFA), mefenamic acid (MFA), and naproxen (NAP) were purchased from Sigma-Aldrich (Saint Louis, MO, USA). Ethyl acetate (99%) was supplied from Penta (Livingston, NJ, USA). All derivatization reagents; *N,O*-Bis(trimethylsilyl)trifluoroacetamide with 1% trimethylchlorosilane (BSTFA & 1% TMCS), *N*-tert-Butyldimethylsilyl-*N*-methyltrifluoroacetamide (MTBSTFA), Pentafluoropropionic anhydride (PFPA), 2,2,3,3,3-Pentafluoro-1-propanol (PFPOH), Trifluoroacetic anhydride (TFAA), Heptafluorobutyric anhydride (HFBA), were for GC derivatization $\geq$ 99% grade and were purchased from Sigma-Aldrich (Saint Louis, MO, USA). Serum samples were obtained from 4 cases that arrived in AHEPA University General Hospital and were suspected for intoxication. Drug-free human serum was obtained from healthy donors and before its use it was screened by GC/MS for the presence of the NSAIDs.

2.2. Preparation of the Standard Solutions

Stock solutions of all compounds were prepared in ACN at 10,000 μg/mL. From these, a mix solution containing all nine drugs was prepared and diluted to the following concentrations: 1000 μg/mL, 200 μg/mL, 100 μg/mL, 40 μg/mL, 20 μg/mL, 4 μg/mL, and 2 μg/mL.

2.3. *Sample Preparation*

For the selection of the optimum pH conditions, 200 μL of spiked human serum at 200 μg/mL was adjusted at pH of 3.7, 4.7, and 5.7 with the addition of 100 μL of HCOOH/HCOONa buffer solutions prepared at these pH values (by mixing appropriate values of 0.2 M solutions of HCOOH and HCOONa). Then 1 mL of ethyl acetate was added and the mixture was shaken for 10 min. After centrifugation for 5 min at 6300× *g* the organic phase was collected and dried under gentle nitrogen stream at room temperature. The obtained residue was redissolved either in 50 μL of ethyl acetate or in 50 μL of derivatization reagents, as described in Section 2.5. In any case, 1 μL of the extract was injected in the GC-MS system.

2.4. *Derivatization Procedure*

For the selection of the optimum derivatization procedure, five different reagents were used: (a) BSTFA with 1% TMCS, (b) MTBSTFA, (c) HFBA, (d) PFPA with PFPOH, and (e)TFAA. The procedure followed for silylation was as follows; addition of 50 μL BSTFA, 1% TMCS, or 50 μL MTBSTFA in the dry residue and after vortexing the mixture was heated for 20 min at 70 °C, finally 1 μL was injected in the system. Acetylation was performed with the addition of either 50 μL of HFBA, or of 30 μL PFPA with 20 μL PFPOH, or of 50 μL of TFAA. The mixture was heated for 20 min at 70 °C and then after cooling down was evaporated and the residue was dissolved in 50 μL of ethyl acetate. From this extract, 1 μL was injected to GC-MS.

2.5. *GC/MS Analysis*

GC-MS analysis was performed on an Agilent Technologies 7890A GC, equipped with a CTC autosampler and combined with a 5975C inert XL EI/CI MSD with Triple-Axis Detector (Agilent Technologies, Santa Clara, CA, USA). GC separations were performed on a 30 m Agilent J&W HP-5ms UI capillary column, with a film thickness of 0.25 μm and an i.d. of 0.25 μm. Back-flash was performed with a 1.5 m deactivated Agilent column with a film thickness of 0.18 mm. The method had a duration of 15 min with the following temperature program: Initial oven temperature at 120 °C, hold for 1 min, and then increase to 300 °C with a 15 °C/min rate. A back-flash step followed, at 300 °C for 10 min. Injection of 1 μL of sample was made on a PTV injector operating from 200 °C to 320 °C. The mass spectrometer (MS) was operated at electron impact ionization mode (EI, 70 eV) and the mass scan range was from 40 to 500 amu.

3. Results

The majority of the studied drugs are categorized in the acidic class of compounds as they contain carboxyl moieties, except from APAP and NI. Their chemical structures, together with their pKa values, can be seen in Table 1. These are expected to be extracted more efficiently by the organic solvent under acidic conditions, where the ionization of their carboxyl group is suppressed. However, as their physicochemical properties are varying due to their structure and the presence of different substitution groups, the optimum pH for their extraction is needed to be examined.

Table 1. Chemical structures and pKa values of the studied non-steroidal anti-inflammatory drugs (NSAIDs).

Compound	Chemical Structure	pKa
APAP	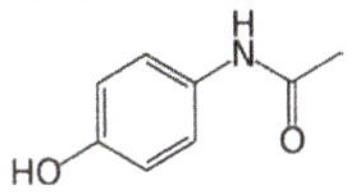 H N O HO	9.38

Table 1. *Cont.*

Compound	Chemical Structure	pKa
DCF	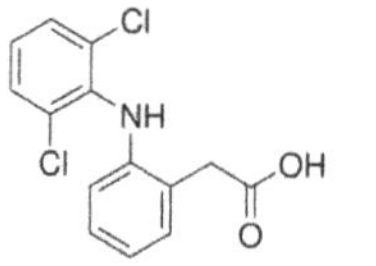	4.15
IBP	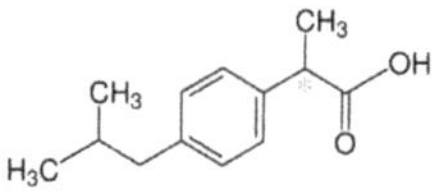	4.91
NI	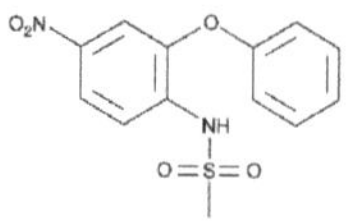	6.86
NFA	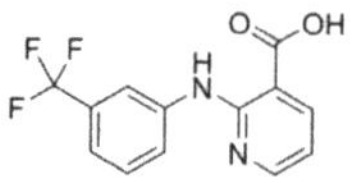	1.88
MFA	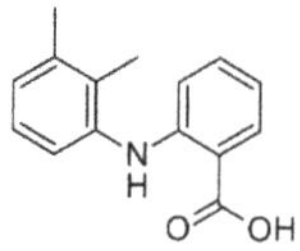	4.2
NAP	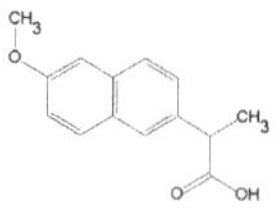	4.15
ASA	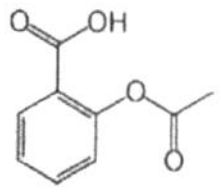	3.49
SA	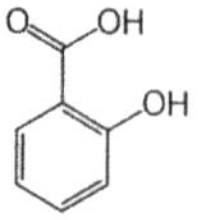	2.97

3.1. Extraction pH

Under three different pH conditions, extraction of the eight NSAIDs from a serum sample spiked at 200 μg/mL was performed, and the extracts were thereafter analyzed without any derivatization. As expected, detection sensitivity of the underivatized drugs is low, however it was acceptable for comparative purposes. Apart from the eight drugs, two more peaks were considered: (1) The peak which corresponds to salicylic acid, a product occurring from ASA in the sample, and (2) the diclofenac-lactam peak, which is obtained by DCF partial conversion during GC analysis. Table 2 provided the peak areas of the studied drugs at three different pHs tested for their extraction. As can be seen, more acidic pH conditions favor the extraction of the majority of the drugs resulting in higher detected peak areas. NI, however, provides higher signal under pH 4.7. For the case of diclofenac, its lactam is detected in higher proportion under all three pH conditions. Apart from that, the methylester of dichlofenac (DCF-ME) is also detected with relatively high signal. Salicylic acid is detected with higher amount under the more acidic pHs compared to acetyl salicylic acid.

Table 2. Absolute area of each NSAID at different pH extraction conditions.

Compound	Absolute Area (10^6)		
	pH = 3.7	pH = 4.7	pH = 5.7
APAP	282.2	261.5	139.0
DCF-lactam/DCF-ME/DCF	78.3/52/24	56.1/27/35	10.8/3.9/-
NAP	28.2	6.2	0.8
NFA	12.0	3.4	0.5
ASA	4.1	1	0.2
SA	10.1	1.1	0.2
IBP	267.6	158.6	20.4
MFA	77.7	74.6	4.4
NI	123.7	185.6	23.6

3.2. *Analysis Without Derivatization*

When the extracted NSAIDs were analyzed directly without prior derivatization, the detection sensitivity was not satisfactory for the majority of the eight drugs. The chromatographic peaks were broad with excessive tailing, which was attributed to intramolecular interactions and the interaction of polar carboxylic groups of drugs with the column's supporting material. Detection of such compounds can be challenging as several transformation products of the drugs were also detected. During GC/MS analysis, NFA, MFA, and NAP were not found stable without derivatization, as their decarboxylated compounds (-CO_2) were also detected. This has also been reported in previous studies [15]. As an example, together with NAP detected at 9.86 min, its NAP-CO_2 degradation product is also detected at 7.04 min with characteristic ions at m/z 184 and 141. Another issue is that ASA is converted to SA over time in both aqueous and organic solutions, thus it can also be detected due to that reason. This can mislead its determination, due to the fact that SA is always detected in serum after administration as it is the active metabolite of ASA. Because of that, freshly prepared solutions of ASA should always be used, and SA should also be determined when ASA is present in serum.

In addition, for the cases of ASA, SA, MFA, NFA, IBP, NAP, and DCF, their methyl ester products were also detected in the spiked extract.

As it concerns DCF, it was observed that when the analysis was performed directly in the serum extract without prior derivatization, only a small amount of DCF could be detected, whereas the highest amount was detected as diclofenac lactam. This was firstly reported by El Haj et al. in 1999, who studied the methanolic solutions of DCF [15]. The authors attributed the formation of lactam to the high temperatures applied in the inlet during the GC-MS analysis. Here, in order to investigate this phenomenon, various analyses were conducted and the findings are discussed below.

3.3. *Derivatization Study*

In order to enhance the chromatographic peak characteristics of the compounds and the detection sensitivity, the derivatization procedure was performed with different reagents. The aim was to select the optimum procedure and facilitate the simultaneous determination of the eight NSAIDs. In spiked serum samples at 200 µg/mL the derivatization procedures were conducted as described in Section 2.4, and the obtained chromatographic traces with the five reagents are summarized in Table 3. The chromatographic peaks were assigned based on the GC-MS libraries used [NIST v11 and Mass Spectral Library of Drugs, Poisons, Pesticides, Pollutants, and Their Metabolites, 3th Edition (PMW_Tox3.l)]. All the obtained peaks could be identified, apart from the acetyl derivatives, the spectra of which didn't exist in any of the used MS libraries, web-based libraries, or in the literature. In Table 3, the retention time and the characteristic ions of each drug or drug derivative are juxtaposed for the five derivatization procedures and that without any derivatization. As can be seen, AS and ASA, could not be detected at all with HFBA, TFA, or PFPA-PFPOH derivatization. In some cases, more than one derivative was detected, for example for APAP with MTBSTFA, while for others the underivatized drug was also detected, such as for NI with silylation reagents. In the case of PFPA-PFPOH, derivatization of NFA gives two peaks which cannot be assigned based on their spectra, as acetyl-derivatives are not registered in the MS spectral libraries used. The major peak is that at 6.9 min, which was attributed to NFA-pfp, while the other peak at 8 min can be another derivative of NFA. Ideally, the optimum procedure should lead to a sole derivative, whereas when the underivatized compound or more than one derivative is detected, the complexity of the detection is increased and the reproducibility of the obtained results is hindered. Based on the obtained results, HFBA, TFA, and PFPA-PFPOH are not considered to be the most appropriate for the simultaneous detection and determination of the eight NSAIDs for the above mentioned reasons.

As it concerns the detection sensitivity of the tested derivatization protocols, silylation provided better results in comparison to the other reagents for all NSAIDs except from IBP. The latter formed a derivative with PFPA/PFPOH which exhibited higher peak area when compared to the BSTFA and MTBSTFA derivatives. Between the two silylation reagents, BSTFA was selected as the best one. In Figure 1, a characteristic total ion chromatogram of serum sample spiked with the studied NSAIDs is presented. The selection was based on the fact that the majority of the peaks had higher peak areas and that MTBSTFA, in the case of APAP, forms two derivatives at the same peak height. Only NI derivatization with BSTFA seems to have low yield, as it gives a small peak of the derivative and the largest peak as NI. However, this is observed for the other silylation reagent as well, whereas NI does not derivatized at all with HFBA, PFPA/PFPOH, and TFA. This means that NI will finally be determined by considering the peak of the underivatized drug. In Table 4, the obtained peak areas are given for the eight drugs, where the higher peak areas can be seen for the silyl-derivatives. In the cases where the peak of the underivatized drug is detected this is noted by an asterisk.

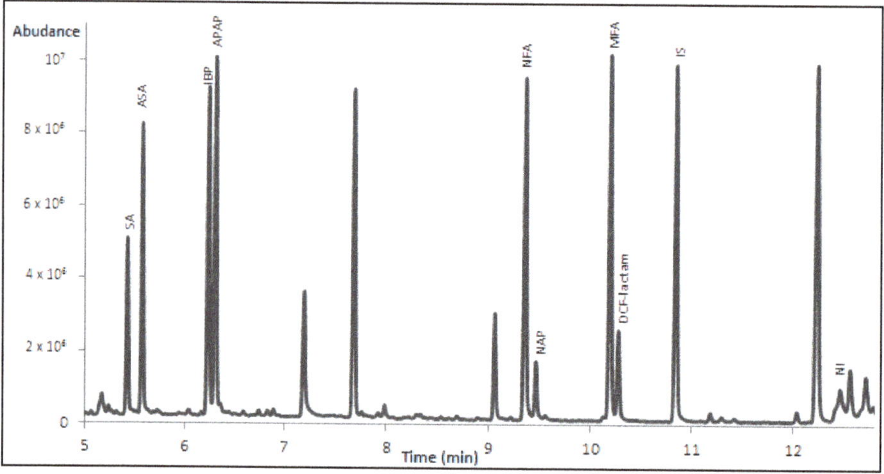

Figure 1. TIC of a serum sample spiked with the studied non-steroidal anti-inflammatory drugs (NSAIDs) at 5 µg/mL, derivatized with BSTFA & 1% TMCS.

Table 3. Retention time, quantifier ion, and two qualifier ions monitored for each NSAID and their derivatives.

	No Derivatization	BSTFA, 1% TMCS	MTBSTFA	HFBA	TFAA	PFPA-PFPOH
Compound	R.T./**Qf**, Ql ions	R.T./**Qf**, Ql ions/der	R.T./**Qf**, Ql ions/der	R.T./**Qf**, Ql ions/der	R.T./**Qf**, Ql ions/der	R.T./**Qf**, Ql ions/der
ASA	4.09/**120**,138,92	5.55/**195**,120/210/ASA-tms	7.20/**195**,237/135/ASA-tbdms	-	-	-
SA	5.24/**120**,92,138	5.48/**267**,135,91/SA-2tms	8.61/**309**,195,209/SA-2tbdms	-	-	-
IBP	6.02/**161**,163,206	6.22/**160**,263,353/IBP-tms	7.85/**263**,75,117/IBP-tbdms	6.02/161,163,206/IBP	7.87/**263**,161,117/IBP-tfa 6.02/161,163,206/IBP	4.99/**161**,295,239/IBP-pfp
APAP	6.89/**109**,151,80	6.29/**206**,280,295/APAP-2tms	8.91/**208**,265/166/APAP-tbdms 9.49/**322**,308,248/APAP-2tbdms	6.16/**108**,305,347/APAP-hfb 4.31/**304**,109,322/no ID *	5.81/**108**,205,247/APAP-tfa	5.89/**108**,255,297/APAP-pfp
NFA	9.10/**237**,263,282	9.34/**236**,263,353/NFA-tms	10.79/**245**,339,265/NFA-tbdms	7.00/**433**,265,168/NFA-hfb	6.90/**333**,265,168/NFA-tfa	6.92/**236**,264,441/NFA-pfp 8.00/**413**,263,236/no ID *
MFA	10.32/**223**,208,241	10.17/**223**,313,208/MFA-tms	11.60/**224**,209,298/MFA-tbdms	8.10/**224**,393,209/MFA-hfb	9.83/**222**,319,250/MFA-tfa 8.03/**294**,224,209/no ID *	8.60/**398**,250,222/MFA-pfp
DCF	10.36/**214**,242,277/DCFlactam 13.01/**214**,242,295/DCF	10.30/**349**,190,314/DCF-lactam-tms 10.82/**214**,242,367/DCF-tms	11.62/**300**,302,391/DCF-lactam-tbdms 12.24/**352**,214,409/DCF-tbdms	10.36/**214**,242,277/DCFlactam	10.36/**214**,242,277/DCF-lactam	10.36/**214**,242,277/DCF-lactam
NAP	9.86/**185**,230,170	9.46/**185**,243,302/NAP-tms	10.84/**287**,185,141/NAP-tbdms	9.86/185,230,170/NAP 7.04/184,141,169/NAP-CO_2	9.86/185,230,170/NAP 7.04/184,141,169/NAP-CO_2	8.24/**185**,170,141/NAP-pfp
NI	12.46/**154**,229,308	12.52/**380**,365,228/NI-tms 12.46/**154**,229,308/NI	12.52/**287**,241,344/NI-tbdms 12.46/**154**,229,308/NI	12.46/**154**,229,308/NI	12.46/**154**,229,308/NI	12.46/**154**,229,308/NI

Qf: Quantifier, Ql: Qualifier, der: Derivative. * Unidentified peak which corresponds to a derivative.

Table 4. Peak areas of the obtained derivatives with the five derivatization protocols applied in the eight NSAIDs.

Compound	Peak Area 10^6				
	BSTFA	MTBSTFA	HFBA	PFPA-PFPOH	TFAA
APAP	330	194/190	190	171	90
IBP	362	266	137 *	455	20.7/219 *
MFA	443	350	188	244	153
NFA	381	350	97.5	310/12 *	22.4
NAP	196	190	8 */4	185	8 */3
ASA	425	420	-	-	-
SA	350	345	-	-	-
NI	115 */2.5	107 */3	130 *	140 *	100 *
DCF-lactam	440	402	306 *	280 *	322 *

* Underivatized compound.

3.4. Application of the Method

With the aim to apply the method to real clinical human samples, recovery of the optimum procedure was examined and linearity and limits of detection and quantification of the method was assessed. Recovery (R%) was experimentally calculated and expressed as the percentage ratio of the peak areas of the serum spiked before extraction at 10 μg/mL to the peak areas of the serum extract spiked after extraction. The R% ranged from 51.50% for ASA to 85.81% for NAP. The R% for all the studied drugs are presented in Table 5.

Table 5. Recovery % of the studied NSAIDs.

Compound	Recovery % (before LLE/after LLE)
APAP	61.35
ASA	51.50
DCF	70.30
IBP	63.17
MFA	64.27
NAP	85.81
NFA	65.15
NI	52.90
SA	59.28

Linearity of the method was evaluated by analyzing human serum samples spiked with a mixture of NSAIDs at concentrations of 0.5, 1, 5, 10, 25, and 50 μg/mL using as internal standard 20 μL nordiazepm-D_5 (C = 5 μg/mL). BSTFA derivatization was applied with 1% TMCS as described in Sections 2.3 and 2.4, and the peak areas ratios were considered for quantitation. For the case of NI, the peak of underivatized drug was considered, as in low concentrations the derivative is not detected at all. The limits of quantitation (LOQ) were experimentally calculated as signal to noise ratio 10:1 and the limits of detection, and limits of detection (LOD) as signal to noise 3:1. The equations of calibration curves based on linear regression, LOQs, and LODs for all drugs are given in Table 6.

Selectivity was determined on the drug-free blood serum samples obtained from six different healthy volunteers from the laboratory staff. No traces of the studied NSAIDs or other interferences could be detected.

Here it should be noted that freshly prepared ASA solutions in ACN were used for the determination of ASA, as it was observed that ASA transforms to SA. This has been observed to proceed faster in methanol than in ACN [16]. As SA is also present in real samples, its co-determination is required for both reasons.

Table 6. Linearity features of the method such as R^2, limits of detection and quantitation, and the linear equation for all drugs.

Compounds	Linear Equation	R^2	LOD (ng/mL)	LOQ (ng/mL)
ASA-tms	y = 0.578x + 0.407	0.994	10	30
SA-2tms	y = 0.456x + 0.965	0.998	5	15
APAP-2tms	y = 0.8222x + 0.8157	0.997	2	6
DCF-lactam-tms	y = 0.3448x + 0.7265	0.991	3	10
MFA-tms	y = 0.4625x + 0.9759	0.993	12	40
NFA-tms	y = 0.516x + 1.0058	0.992	13	43
IBP-tms	y = 0.567x + 1.7578	0.993	2	6
NAP-tms	y = 0.3187x + 1.1305	0.996	10	34
NI *	y = 0.7793x − 1.4552	0.999	124	414

* Nimesulide was determined underivatized.

The accuracy and precision of the method was evaluated within a day (n = 4) and over a period of a week (n = 3) at low, medium, and high concentration levels (1, 10, 50 μg/mL). The intra-day accuracy and precision were found to be between 1.03%–9.79% and 88%–110%, respectively, while the inter-day accuracy and precision were between 1.87%–10.79% and 91%–113%. In Table 7, the data from accuracy and precision assays are presented.

Table 7. Intra- and inter-assay data of the method.

		Intra-Assay (n = 4)				Inter-Assay (n = 3)			
Compound	Added (μg/mL)	Mean Found (μg/mL)	SD	CV %	Accuracy %	Overall Mean Found (μg/mL)	SD	CV %	Accuracy %
ASA	1	1.09	0.02	1.83	109.00	1.07	0.02	1.87	107.00
	10	11.00	0.60	5.45	110.00	11.30	0.61	5.40	113.00
	50	48.00	1.81	3.77	96.00	49.10	2.10	4.28	98.20
SA	1	0.98	0.05	5.10	98.00	0.99	0.06	6.06	99.00
	10	9.85	0.51	5.18	98.50	9.87	0.54	5.47	98.70
	50	51.10	1.20	2.35	102.20	50.80	1.30	2.56	101.60
APAP	1	0.97	0.01	1.03	97.00	1.00	0.03	3.00	100.00
	10	10.70	0.30	2.80	107.00	10.40	0.32	3.08	104.00
	50	49.80	0.90	1.81	99.60	50.20	1.20	2.39	100.40
DCF	1	0.88	0.04	4.55	88.00	0.91	0.06	6.59	91.00
	10	10.62	0.24	2.26	106.20	10.42	0.35	3.36	104.20
	50	49.20	1.50	3.05	98.40	49.30	1.90	3.85	98.60
MFA	1	1.12	0.05	4.46	112.00	1.05	0.06	5.71	105.00
	10	9.30	0.60	6.45	93.00	9.70	0.63	6.49	97.00
	50	51.30	2.40	4.68	102.60	50.90	3.10	6.09	101.80
NFA	1	0.98	0.03	3.06	98.00	1.01	0.05	4.95	101.00
	10	10.20	0.46	4.51	102.00	10.30	0.64	6.21	103.00
	50	51.60	3.20	6.20	103.20	51.60	3.50	6.78	103.20
IBP	1	0.94	0.07	7.45	94.00	0.96	0.08	8.33	96.00
	10	9.76	0.75	7.68	97.60	9.83	0.89	9.05	98.30
	50	49.80	2.60	5.22	99.60	50.30	3.20	6.36	100.60
NAP	1	1.02	0.05	4.90	102.00	1.06	0.08	7.55	106.00
	10	9.79	0.46	4.70	97.90	9.86	0.53	5.38	98.60
	50	50.50	4.60	9.11	101.00	50.81	2.26	4.45	101.62
NI	1	0.95	0.06	6.32	95.00	0.98	0.07	7.14	98.00
	10	9.86	0.42	4.26	98.60	9.88	0.56	5.67	98.80
	50	52.10	5.10	9.79	104.20	51.90	5.60	10.79	103.80

The method was then applied to four clinical cases where exposure to NSAIDs was suspected. The findings are presented in Table 8. S1 and S2 correspond to serum samples which were taken 12 h after APAP ingestion, and they were found below toxic levels (below 50 μg/mL, Rumack-Matthew nomogram). S3 corresponds to a case of a 13 year old girl claiming a suicide attempt with DCF, however the concentration after 3 h at 2.31 μg/mL was at the therapeutic levels. S4 corresponds to a patient treated with ASA (the sample was taken 40 min after ingestion) and the concentration was found to be below the therapeutic levels. The extracted ion chromatograms of S2 and S3 are presented in Figure 2a,b.

Table 8. Concentrations found with the applied method in the serum cases positive to NSAIDs.

Sample/Compound Found	Concentration Found (μg/mL) ± SD
S1/APAP	18.2 ± 0.3
S2/APAP	7.50 ± 0.02
S3/DCF	2.31 ± 0.05
S4/ASA	2.0 ± 0.2
S4/SA	0.21 ± 0.02

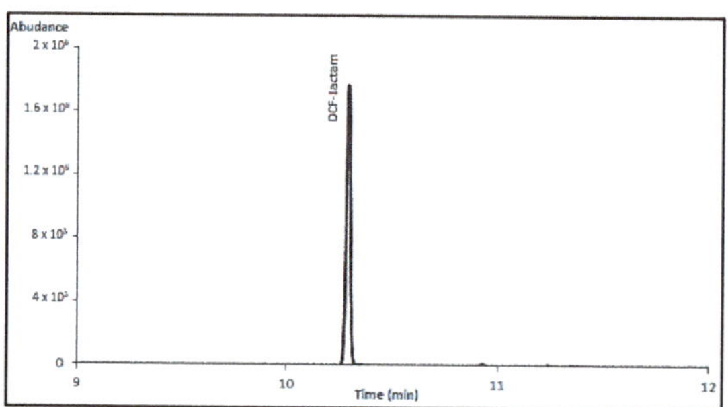

(**a**)

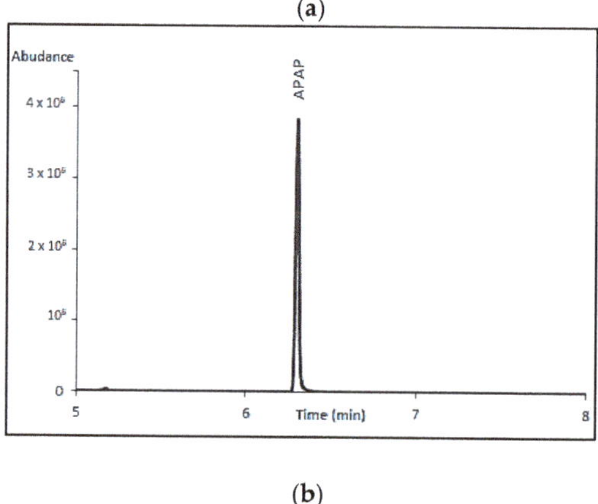

(**b**)

Figure 2. Extracted ion chromatograms of real serum sample (**a**) S3 at 349 m/z found positive for diclofenac determined at 2.31 μg/mL and (**b**) S2 at 206 m/z found positive for APAP determined at 7.5 μg/mL.

4. Discussion

According to our findings, more efficient extraction of the studied drugs was performed under an acidic pH of 3.7. NI showed also high extraction recovery in higher pH as well.

For most of the studied drugs, derivatization was required for their sensitive detection, however NI provided high, sharp peak in the initial extract and can be best determined without any derivatization.

Some of our findings lead us to the conclusion that determination of these drugs needs cautious sample treatment and data interpretation. We have observed, like previous findings [16], that ASA is not stable in solution and it is hydrolysed into SA, and finally over time both compounds are in equilibrium in the solution. Degradation of ASA has been studied thoroughly [16–18], however there are recent studies where the authors overlook this parameter [1,19]. In order to overcome this transformation, freshly prepared solution should be used for ASA.

Without derivatization, the majority of the NSAIDs seem to be unstable under the high temperature conditions in GC-MS, and a variety of different derivatives are also detected.

The decarboxylated products of NAP, MFA, and NFA are detected to elute some minutes earlier than the parent compound peaks. The formation of these compounds have been reported by others [15] and most probably is due to the fact that the underivatized compounds are labile compared to their derivatives under the same high temperature conditions in GC-MS.

In addition, for almost all the studied drugs, their methylesters were also detected. The esterification process seems to take place at the high temperature conditions in GC-MS, especially in methanolic solutions, whereas this doesn't seem to happen for the derivatives. For this, ACN was used as a solvent.

What is most interesting is that DCF transformed to its dehydrated product, the lactam. To investigate this, a series of experiments was conducted. First, a solution of DCF was analyzed directly with GC-MS, and it was observed that almost all of the diclofenac was transformed to its lactam form. When the same solution was analyzed after derivatization with BSTFA + 1% TMCS only diclofenac-tms was determined, indicating stability of the molecule with derivatization. Then a real human serum sample positive to DCF and a human serum sample spiked with DCF were analyzed after derivatization with BSTFA + 1% TMCS. In both cases, and in contrast to the previous finding, DCF-lactam-tms was mainly detected. Contrastingly, in an aqueous solution of DCF, which was analyzed after extraction and derivatization, similarly to a real sample, only DCF-tms could be detected. This could mean that diclofenac-tms is more stable than the underivatized drug, which is dehydrated under high temperatures in the GC-MS. However, it seems that the serum matrix under acidic conditions enhances the full conversion of DCF to lactam, as DCF-lactam-tms is the sole peak detected in this case, whereas in the absence of serum matrix DCF-tms is detected. The transformation of DCF to DCF-lactam in water samples, pharmaceutical dosage forms, and urine was reported previously but it hasn't been thoroughly studied [11,16]. In a study, DCF-lactam was falsely reported as DCF [20], and in another study DCF-lactam-tms wasn't detected at all [21].

5. Conclusions

Based on our findings, the most efficient sample preparation protocol for the accurate determination of the studied commonly prescribed NSAIDs is derivatization, more specifically with BSTFA, except from NI where no derivatization step is needed, after acidic extraction.

For some of these drugs, cautious handling is needed, as ASA hydrolyses quickly in solution and DCF is converted to its lactam form in the serum matrix under acidic pH. This conversion is enhanced at high temperatures when its carboxyl group is not protected.

The method needs only 200 μL of blood serum and can determine even trace amounts of the studied compounds.

Author Contributions: Conceptualization, A.K. and H.G.; Methodology, A.K. and N.R.; Validation, A.K. and E.T.; Formal Analysis, A.K. and E.M.; Resources, N.R.; Data Curation, A.K. and E.M.; Writing-Original Draft Preparation, A.K. and H.G.; Writing-Review & Editing, G.T.; Supervision, G.T. and H.G.

Conflicts of Interest: The authors declare no conflict of interest.

Abbreviations

ASA	Acetyl salicylic acid
SA	Salicylic acid
NAP	Naproxen
IBP	Ibuprofen
APAP	Acetaminophen
NFA	Niflumic acid
MFA	Mefenamic acid
NI	Nimesulide
DCF	Diclofenac

References

1. Sirok, D.; Pátfalusi, M.; Szeleczky, G.; Somorjai, G.; Greskovits, D.; Monostory, K. Robust and sensitive LC/MS-MS method for simultaneous detection of acetylsalicylic acid and salicylic acid in human plasma. *Microchem. J.* **2018**, *136*, 200–208. [CrossRef]
2. Bylda, C.; Thiele, R.; Kobold, U.; Volmer, D.A. Simultaneous quantification of acetaminophen and structurally related compounds in human serum and plasma. *Drug Test. Anal.* **2014**, *6*, 451–460. [CrossRef] [PubMed]
3. Yilmaz, B.; Sahin, H.; Erdem, A.F. Determination of naproxen in human plasma by GC-MS. *J. Sep. Sci.* **2014**, *37*, 997–1003. [CrossRef] [PubMed]
4. Vinci, F.; Fabbrocino, S.; Fiori, M.; Serpe, L.; Gallo, P. Determination of fourteen non-steroidal anti-inflammatory drugs in animal serum and plasma by liquid chromatography/mass spectrometry. *Rapid Commun. Mass Spectrom.* **2006**, *20*, 3412–3420. [CrossRef] [PubMed]
5. Sun, X.; Xue, K.L.; Jiao, X.Y.; Chen, Q.; Xu, L.; Zheng, H.; Ding, Y.F. Simultaneous determination of nimesulide and its four possible metabolites in human plasma by LC-MS/MS and its application in a study of pharmacokinetics. *J. Chromatogr. B* **2016**, *1027*, 139–148. [CrossRef] [PubMed]
6. Taylor, R.R.; Hoffman, K.L.; Schniedewind, B.; Clavijo, C.; Galinkin, J.L.; Christians, U. Comparison of the quantification of acetaminophen in plasma, cerebrospinal fluid and dried blood spots using high-performance liquid chromatography-tandem mass spectrometry. *J. Pharm. Biomed. Anal.* **2013**, *83*, 1–9. [CrossRef] [PubMed]
7. Chhonker, Y.S.; Pandey, C.P.; Chandasana, H.; Laxman, T.S.; Prasad, Y.D.; Narain, V.S.; Dikshit, M.; Bhatta, R.S. Simultaneous quantitation of acetylsalicylic acid and clopidogrel along with their metabolites in human plasma using liquid chromatography tandem mass spectrometry. *Biomed. Chromatogr.* **2016**, *30*, 466–473. [CrossRef] [PubMed]
8. Hložek, T.; Bursová, M.; Čabala, R. Fast ibuprofen, ketoprofen and naproxen simultaneous determination in human serum for clinical toxicology by GC–FID. *Clin. Biochem.* **2014**, *47*, 109–111. [CrossRef] [PubMed]
9. Chang, K.C.; Lin, J.S.; Cheng, C. Online eluent-switching technique coupled anion-exchange liquid chromatography–ion trap tandem mass spectrometry for analysis of non-steroidal anti-inflammatory drugs in pig serum. *J. Chromatogr. A* **2015**, *1422*, 222–229. [CrossRef] [PubMed]
10. Elsinghorst, P.W.; Kinzig, M.; Rodamer, M.; Holzgrabe, U.; Sörgel, F. An LC-MS/MS procedure for the quantification of naproxen in human plasma: Development, validation, comparison with other methods, and application to a pharmacokinetic study. *J. Chromatogr. B* **2011**, *879*, 1686–1696. [CrossRef] [PubMed]
11. Ding, Y.; Garcia, C.D. Determination of Nonsteroidal Anti-inflammatory Drugs in Serum by Microchip Capillary Electrophoresis with Electrochemical Detection. *Electroanalysis* **2006**, *18*, 2202–2209. [CrossRef]
12. Payán, M.R.; López, M.Á.B.; Fernández-Torres, R.; Bernal, J.L.P.; Mochón, M.C. HPLC determination of ibuprofen, diclofenac and salicylic acid using hollow fiber-based liquid phase microextraction (HF-LPME). *Anal. Chim. Acta* **2009**, *653*, 184–190.
13. Way, B.A.; Wilhite, T.R.; Smith, C.H.; Landt, M. Measurement of plasma ibuprofen by gas chromatography-mass spectrometry. *J. Clin. Lab. Anal.* **1997**, *11*, 336–339. [CrossRef]
14. Bhushan, R.; Joshi, S.; Arora, M.; Gupta, M. Study of the liquid chromatographic separation and determination of NSAID. *JPC J. Planar Chromatogr. Mod. TLC* **2005**, *18*, 164–166. [CrossRef]

15. El Haj, B.M.; Al Ainri, A.M.; Hassan, M.H.; Bin Khadem, R.K.; Marzouq, M.S. The GC/MS analysis of some commonly used non-steriodal anti-inflammatory drugs (NSAIDs) in pharmaceutical dosage forms and in urine. *Forensic Sci. Int.* **1999**, *105*, 141–153. [CrossRef]
16. Skibinski, R.; Komsta, L. The stability and degradation kinetics of acetylsalicylic acid in different organic solutions revisited—An UHPLC-ESI-QTOF spectrometry study. *Curr. Issues Pharm. Med. Sci.* **2016**, *29*, 39–41. [CrossRef]
17. Marra, M.C.; Cunha, R.R.; Vidal, D.T.; Munoz, R.A.; do Lago, C.L.; Richter, E.M. Ultra-fast determination of caffeine, dipyrone, and acetylsalicylic acid by capillary electrophoresis with capacitively coupled contactless conductivity detection and identification of degradation products. *J. Chromatogr. A* **2014**, *1327*, 149–154. [CrossRef] [PubMed]
18. Abuirjeie, M.A.; Abdel-hamid, M.E.; Ibrahim, E.-S.A. Simultaneous High-Performance Liquid Chromatographic Assay of Acetaminophen, Acetylsalicylic Acid, Caffeine, and D-Propoxyphene Hydrochloride. *Anal. Lett.* **1989**, *22*, 365–375. [CrossRef]
19. Kees, F.; Jehnich, D.; Grobecker, H. Simultaneous determination of acetylsalicylic acid and salicylic acid in human plasma by high-performance liquid chromatography. *J. Chromatogr. B* **1996**, *677*, 172–177. [CrossRef]
20. Yilmaz, B.; Ciltas, U. Determination of diclofenac in pharmaceutical preparations by voltammetry and gas chromatography methods. *J. Pharm. Anal.* **2015**, *5*, 153–160. [CrossRef] [PubMed]
21. Yilmaz, B. GC-MS Determination of Diclofenac in Human Plasma. *Chromatographia* **2010**, *71*, 549–551. [CrossRef]

Article

Separation Optimization of a Mixture of Ionized and Non-Ionized Solutes under Isocratic and Gradient Conditions in Reversed-Phase HPLC by Means of Microsoft Excel Spreadsheets

Chrysostomi Zisi, Athina Maria Mangipa, Eleftheria Boutou and Adriani Pappa-Louisi *

Department of Chemistry, Aristotle University of Thessaloniki, 54124 Thessaloniki, Greece; chrizisi@hotmail.com (C.Z.); amangipa@chem.auth.gr (A.M.M.); empoutou@chem.auth.gr (E.B.)

* Correspondence: apappa@chem.auth.gr; Tel.: +30-2310-997765

Received: 2 February 2018; Accepted: 12 March 2018; Published: 18 March 2018

Abstract: The crucial role of mobile phase pH for optimizing the separation of a mixture of ionized and non-ionized compounds on a Phenomenex extended pH-range reversed-phase column (Kinetex 5 μm EVO C18) was examined. A previously developed Excel-spreadsheet-based software was used for the whole separation optimization procedure of the sample of interest under isocratic conditions as well as under single linear organic modifier-gradients in different eluent pHs. The importance and the advantages of performing a computer-aided separation optimization compared with a trial-and-error optimization method were realized. Additionally, this study showed that the optimized separation conditions for a given stationary phase may be used to achieve successful separations on new columns of the same type and size. In general, the results of this work could give chromatographers a feel of confidence to establish desired separations of a mixture of ionizable and neutral compounds in reversed-phase columns.

Keywords: reversed-phase liquid chromatography; ionizable and non-ionizable analytes; isocratic and gradient elution in different eluent pHs; computer-assisted separation optimization; visualization of predicted chromatograms

---

1. Introduction

Reversed-phase liquid chromatography (RPLC) is one of the most widely used chromatographic techniques. It is a consequence of its universality, relatively low costs, and general simplicity of analytical procedures. Still, the development of any method with the desired RPLC separation might be long due to the large number of chromatographic settings that might be adjusted (mobile phase composition, pH, temperature, etc.). The most popular trial-and-error approach has several disadvantages since it is often time-consuming, usually requires a large number of preliminary experiments, and might not be fully efficient. Model-based techniques (fully or semi-automated software programs) can be used in the process of searching for desired RPLC separations [1–5]. These methods usually provide very successful separations based on a series of preliminary experiments. We believe that the number of experimental data may be reduced by utilizing the optimal separation conditions predicted for a specific column by an optimization software to other new columns of the same type and size.

In the present contribution, we report on the optimization of reversed-phase separations of mixture of ionized and non-ionized solutes under isocratic conditions as well as under single linear organic modifier-gradients in different eluent pHs using the Excel-spreadsheet-based software previously developed for simulating and optimizing liquid chromatographic separations [6]. Furthermore, the importance and the advantages of performing a computer-aided separation optimization compared with a trial-and-error optimization method will be confirmed as well as the crucial role of mobile phase pH for optimizing the separation of ionizable compounds [7,8]. This study also provides an example that the optimized separation conditions predicted for a certain column by an optimization software result in almost baseline separation of the test analytes in different columns of the same type and size.

2. Materials and Methods

2.1. Materials and Reagents

All chemicals were used as received from commercial sources. The solutes tested are: four monoprotic acids, 2-bromo-4-nitrophenol (2B-4NP), 4-bromo-2-nitrophenol (4B-2NP), 3-bromophenol (3-BP), and 2,4-dibromophenol (2,4-DBP); two monoprotic bases, p-chloro aniline (p-CA) and p-bromoaniline (p-BA); and two non-ionized compounds, benzene (B) and toluene (T). The orthophosphate system (85% H_3PO_4, KH_2PO_4, Na_2HPO_4) was employed for the preparation of buffer solutions in HPLC studies. Acetonitrile (ACN) or methanol (MeOH) of HPLC grade was used as organic modifiers.

2.2. Buffers and Standard Sample Solutions

Aqueous phosphate buffers with a total ionic strength of 0.2 M were used for preparing the mobile phases with different pH values. The composition of the different buffers employed was found in [9]. The working solutions with single solute or solute mixtures were prepared at a concentration of 360 μg mL^{-1} for benzenes, 240 μg mL^{-1} for 3-BP and 2,4-DBP, 24 μg mL^{-1} for 2B-4NP and 4B-2NP and 8 μg mL^{-1} for anilines.

2.3. HPLC System and Conditions

The liquid chromatography system consisted of a Shimadzu LC-20AD pump, a Shimadzu DGU-20A3 degasser, a model 7125 syringe loading sample injector fitted with a 5 μL loop and a Shimadzu UV–visible spectrophotometric detector (Model SPD-10A, Kyoto, Japan) operating at 254 nm. The column was thermostatted at 25 °C by a CTO-10AS Shimadzu column oven.

Four different Phenomenex reversed phase columns of the same size (150 × 4.6 mm) were used. Three of them (Kinetex 5 μm EVO C18, Kinetex 5 μm XB-C18 and Kinetex 2.6 μm XB-C18) were of core–shell technology i.e., with core–shell silica of different particle sizes. The Kinetex 5 μm EVO C18 column exhibits high pH stability from 1–12 similar with that of fully porous organo-silica column (Gemini 5 μm NX-C18), which was also used in this study.

The systematic chromatographic behavior of solutes was investigated on the Kinetex 5 μm EVO C18 column and in mobile phases consisting of diluted aqueous phosphate buffers with a total ionic strength of 0.02 M and a fixed pH value at 2, 3, 5, 7, or 9 modified with ACN. The mobile phase pHs were measured in aqueous buffers before the addition of the organic solvent. Three isocratic runs were performed in different eluent pHs with different ACN volume fraction, (ϕ = 0.3, 0.4, and 0.5) and three ϕ-gradient runs were performed by linearly increasing the ACN content in the mobile phase from an initial value of volume fraction ϕ_0 = 0.3 to a final one ϕ_f = 0.5. In all gradients, a linear elution program was applied with the same starting time (t_{in} = 0 min) but with different gradient duration, t_G. Moreover, the effect brought on retention of test solutes the use of MeOH as organic modifier instead of ACN was investigated at a fixed eluent $pH = 3$ performing three isocratic runs with ϕ_{MeOH} = 0.4, 0.5 and 0.6 as well as three simple linear ϕ_{MeOH}-gradient runs from ϕ_0 = 0.4 to ϕ_f = 0.6 with different gradient duration. The retention data obtained under the above chromatographic conditions, as well as under optimal conditions determined by the optimization procedure adopted in this study, are given in Table S1 in the Supporting Information.

The hold-up time of the Kinetex 5 μm EVO C18 column was estimated to be t_0 = 0.983 min, whereas the dwell time t_D = 0.73 min, at the flow rate set at 1.0 mL min^{-1}.

2.4. Excel-Spreadsheet-Based Optimization Software

The separation optimization procedure under isocratic and simple gradient conditions in different eluent pHs modified with ACN or MeOH was performed using the Excel-spreadsheet-based program developed for simulating and optimizing liquid chromatographic separations [6]. The Excel-spreadsheet -based software 'Isocr&GradSeparationOptimization', with detailed instructions, available on the ACS Publications website at doi:10.1021/acs.jchemed.7b00108 was modified in order to be used for the separation optimization of the mixture of solutes under consideration. An Excel file with the name "Optimization in eluent pH 3 modified with ACN" is provided in the Supporting Information as an example for the whole computer-aided separation optimization procedure adopted in this study.

3. Results

3.1. Effect of the Eluent pH on the Retention of Ionizable Solutes

The effect of eluent pH on the retention factor, k, of a monoprotic acid ($HA \leftrightarrow H^+ + A^-$) or base ($BH^+ \leftrightarrow H^+ + B$) may be expressed as [10–12]

$$k = \frac{k_0 + k_1 10^{j(pH-pK)}}{1 + 10^{j(pH-pK)}} \tag{1}$$

where k_0 and k_1 are the retention factors of the neutral and fully ionized species of these ionogenic analytes, j is an indicator parameter, which is equal to 1 for acids and -1 for bases and $pK = -logK$, K being the equilibrium constant of the appropriate acid/base equilibrium in the eluent, given by $K = \frac{[H^+][A^-]}{[HA]}$ for a monoprotic acid and by $K = \frac{[H^+][B]}{[BH^+]}$ for a monoprotic base. Note that the values of k_0, k_1, and pK depends on the organic content of mobile phase and can be determined by fitting to Equation (1) either isocratic data obtained in different eluent pHs modified with the same organic content or by fitting gradient data of a fixed change of organic content with a fixed gradient duration in different eluent pHs [13].

The influence of eluent pH on the retention of each of the examined solutes is shown in Figure 1 created by fitting to Equation (1) the experimental retention data obtained under isocratic conditions in different eluent pHs with ϕ_{ACN} = 0.3, which are given in Table S1. From this figure it is clear the influence of pH on the different types of ionogenic analytes, as well as the superiority of eluent pH 3 in the separation of the test mixture of ionized and non-ionized solutes tested.

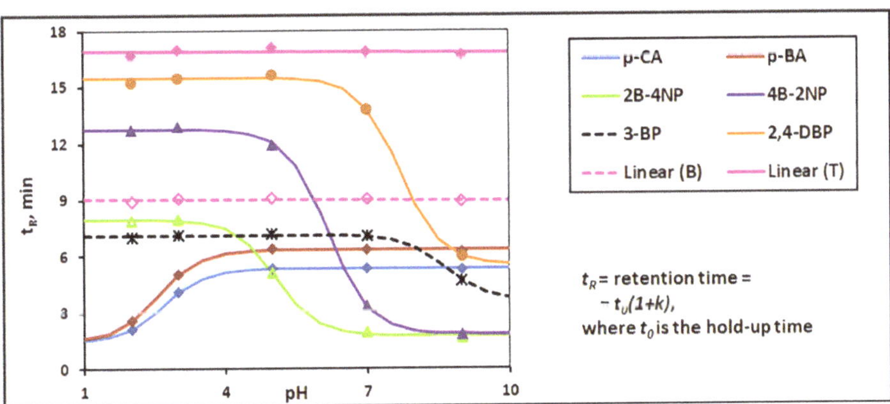

Figure 1. Variation of t_R as a function of mobile phase pH for each of the examined solutes. Points are experimental data taken from Table S1 for isocratic runs performed in different eluent pHs with ϕ_{ACN} = 0.3. Lines are obtained by fitting experimental data of ionized solutes to Equation (1).

3.2. Computer-Aided Separation Optimization in Different Eluent pHs

A computer-assisted separation optimization of a mixture of solutes comprising a visualization of predicted chromatograms involves the following steps [6]:

1. An initial experimental study of the chromatographic behavior of solutes obtained by the least number of chromatographic runs adequately selected.
2. The fitting of the experimental retention data of each solute to a retention model in order to determine its adjustable parameters.
3. The modeling of the peak shape of analytes in the form of a preferred function.
4. The determination of the optimal separation conditions—i.e., the conditions that lead to the best separation of solutes under consideration—based on the above determined retention and peak shape parameters of solutes.
5. The comparison of the simulated chromatogram plotted under the optimal predicted conditions with the corresponding experimental chromatogram recorded under the same conditions in order to test the accuracy of the optimization process.

The whole separation optimization procedure of the test analytes under isocratic conditions as well as under single linear ϕ-gradients in different eluent pHs was implemented on different MS Excel spreadsheets. The optimization procedure followed for separation optimization in eluent pH 3 is given as an example.

The isocratic retention data of solutes, $t_R(exp)$, given in Table S1 and obtained on the Kinetex 5 μm EVO C18 column with different ACN volume fraction (ϕ = 0.3, 0.4, and 0.5) in eluent pH 3 were fitted to the quadratic retention model $\boldsymbol{lnk} = \boldsymbol{c_0 + c_1\varphi + c_2\varphi^2}$ [14] (where k is the solute retention factor, $\boldsymbol{k = (t_R - t_0)/t_0}$, and ϕ is the volume fraction of ACN in the mobile phase) using the spreadsheet 'retention fit' of the file "Optimization in eluent pH 3 modified with ACN" provided in the Supporting Information. Note that, although the spreadsheet 'retention fit' is designed for fitting isocratic retention data to the quadratic retention model, the linear retention model, $\boldsymbol{lnk} = \boldsymbol{c_0 + c_1\varphi}$, could be used instead, if the chromatographic behavior of solutes was studied in a very narrow range of ϕ.

After correction of the baseline of the experimental chromatograms recorded in eluents with ϕ_{ACN} = 0.3 and 0.4 by pressing *Ctrl+q* on the spreadsheet with the name 'BLcor.', available in "Optimization in eluent pH 3 modified with ACN" file, modeling of the peak shapes is straightforward. For fitting the peak shapes of each solute recorded in different chromatograms with ϕ_{ACN} = 0.3 and 0.4 to the model

$$y = h(t_R)exp\left(\frac{-(t - t_R)^2}{D(t_R)^2}\right)$$

the spreadsheet 'peak shape fit' is used. This spreadsheet is designed to estimate the peak shape parameters (h_0, h_1, h_2, D_0, D_1, and D_2) of a quadratic dependence of both peak height, h, and peak width, D, on t_R, given by $\boldsymbol{h(t_R) = h_0 + h_1t_R + h_2t_R^2}$ and $\boldsymbol{D(t_R) = D_0 + D_1t_R + D_2t_R^2}$. It should be noted that the data for two experimental chromatograms are enough for the above procedure since the peaks experimentally recorded in this study permit a linear dependence of both the peak height and peak width on t_R instead of the quadratic one initially assumed.

After the retention times and peak shape parameters for all of the solutes studied under isocratic conditions have been estimated, the values of these parameters (i.e., c_0, c_1, c_2, h_0, h_1, h_2, D_0, D_1, and D_2) are transferred into the spreadsheet 'isocr.optim.' in "Optimization in eluent pH 3 modified with ACN" file. A screenshot of this spreadsheet is displayed in Figure 2. The isocratic separation optimization of the sample mixture is easily automated by pressing *Ctrl+w*. Then, the minimum resolution, R_s, and the maximum of t_R values, $t_R(max)$, of all solutes separated under isocratic conditions in eluent pH 3 are recorded as a function of the organic content ϕ on columns A, B, and C, where ϕ is altered between two values ($\phi(min)$ = 0.3 and $\phi(max)$ = 0.5, preset in cells B15 and B16) with a selected interval $\delta\phi$ = 0.005 placed in cell B17. Simultaneously, the values of R_s and $t_R(max)$ vs. ϕ are plotted in a graph, see the inset

Graph A of the layout of this worksheet, and simulated chromatograms are generated for each mobile phase strength ϕ in the inset Graph B of the worksheet. The execution of the macro is accomplished by finding the optimal eluent, ϕ_{ACN} = 0.365, which leads to the best separation of the sample—i.e., the separation with a desirable value of resolution—R_s = 1.5 (preset in cell D11), in the shortest separation time, which in this case is only 10.36 min. The inset Graph B of Figure 2 depicts a perfect similarity between the simulated chromatogram created for the optimal eluent with ϕ_{ACN} = 0.365 (plotted as the red solid line) and the original experimental one (plotted as the blue dashed line).

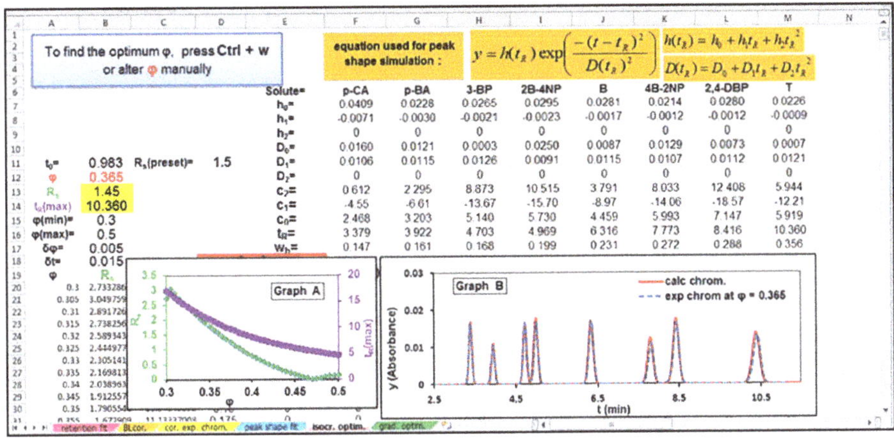

Figure 2. Screenshot of the MS Excel supplementary spreadsheet 'isocr. optim.' used for isocratic separation optimization of solutes in eluent pH 3 modified with ACN. See the text for details.

A procedure similar with that described above for isocratic separation optimization and simulation is also followed for optimizing single linear gradient conditions and simulating chromatograms obtained under selected different gradient profiles. A screenshot of the spreadsheet 'grad. optim.' is depicted in Figure 3.

The values of ϕ_0 and ϕ_f—i.e., ϕ_0 = 0.3 and ϕ_f = 0.5 for the data set analyzed—are placed in cells G2 and G3, respectively, the features of chromatographic system—i.e., the values of t_D = 0.73 min and t_0 = 0.983 min in cells B9 and B10—whereas the estimated retention and peak shape parameters of all solutes are transferred in cells I2:Q10. Note that, in this procedure the retention adjustable parameters, c_0, c_1, and c_2, were determined in the worksheet 'retention fit' from initial isocratic conditions. In contrast, the peak shape parameters, h_0, h_1, h_2, D_0, D_1, and D_2, were obtained from gradient runs between ϕ_0 = 0.3 to ϕ_f = 0.5 with different gradient durations, t_G = 5 and 20 min, in the worksheet 'peak shape fit', since the peak widths in gradient elution are normally compressed compared to those obtained under isocratic conditions. By pressing *Ctrl+e*, the minimum resolution, R_s, and the maximum of t_R values, $t_R(max)$, of all solutes separated under gradient conditions with different gradient durations, t_G, are calculated in columns A, B, and C, where t_G, varied between two values, $t_G(min)$ = 5 min and $t_G(max)$ = 20 min, defined in cells B15 and B16 with a selected interval δt_G − 0.5 min (placed in cell B17). Moreover, a plot is simultaneously created with these values of R_s and $t_R(max)$ vs. t_G, as well as a graph is generated for simulated chromatograms obtained under different gradient profiles. Again, the execution of the macro is accomplished by finding the optimal gradient duration, t_G = 6.5 min, which leads to the best separation of the sample—i.e., the separation with a satisfactory resolution, i.e., R_s = 1.5 (preset in cell D10)—in the shortest separation time, which in this case is only 8.15 min. The inset Graph B of Figure 3 depicts a perfect similarity between the predicted/simulated chromatogram in the optimal gradient elution with a duration t_G = 6.5 min (plotted as the red solid line) and the original experimental one (plotted as the blue dashed line).

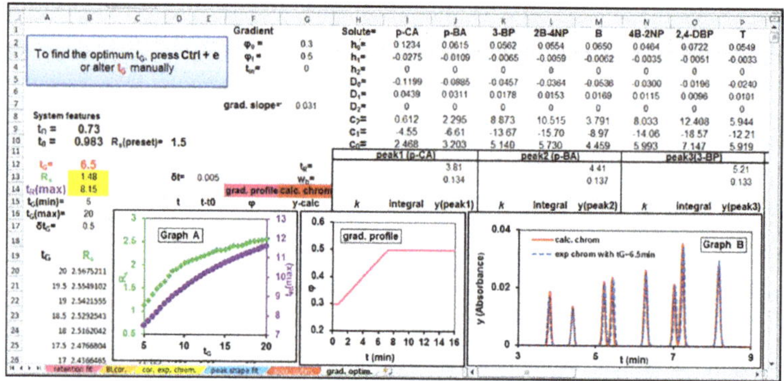

Figure 3. Screenshot of the MS Excel supplementary spreadsheet 'grad. optim.' used for separation optimization under single linear gradient conditions of solutes in eluent pH 3 modified with ACN. See the text for details.

The same optimization approach is applied to chromatographic data obtained under isocratic conditions as well as under single linear ϕ-gradients in other examined eluent pHs—i.e., at pH = 2, 5 and 7, respectively—depicted in Table S1. The retention data recorded in eluent pH = 9 were not analyzed by means of the above spreadsheet optimization program since peak shape distortions appeared for some solutes at that mobile phase pH. The optimal conditions found for the separation of test mixture of solutes on the Kinetex 5 µm EVO C18 column in different eluent pHs are summarized in Table 1. The chromatograms recorded under the optimal gradient conditions found by the optimization algorithm are depicted in Figure 4. In the same Figure, the influence of mobile phase pH on the elution order as well as on the peak shape of ionizable compounds is also illustrated.

Table 1. Optimal conditions found for the separation of the mixture of solutes using an Excel spreadsheet-based optimization program.

pH	Optimum Isocratic Conditions in an Eluent Modified with ACN			Optimum Gradient Conditions for the Linear Variation of ϕ_{ACN} from ϕ_0 = 0.3 to ϕ_f = 0.5		
	ϕ_{ACN}	R_s	$t_R(max)$, min	t_G, min	R_s	$t_R(max)$, min
2	0.370	1.41	9.87	6.5	1.47	8.10
3	0.365	1.45	10.96	6.5	1.48	8.15
5	0.365	1.38	10.40	5.0	1.43	7.46
7	0.370	1.42	10.07	6.0	1.42	7.93

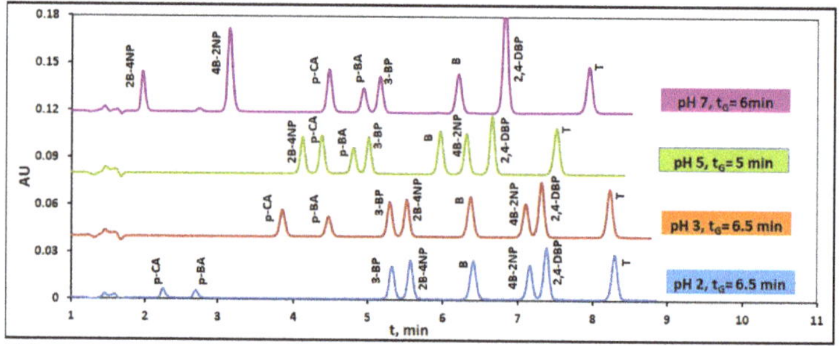

Figure 4. UV detected chromatograms of the mixture of 8 ionized and non-ionized solutes obtained on Kinetex 5 µm EVO column under optimal gradient conditions in different eluent pHs. The elution order of solutes is shown in the Figure. See Table 1 for details of optimal gradient conditions.

3.3. Utility of Computer-Aided Separation Optimization

Once the optimal separation conditions of the sample of interest were found for the Kinetex 5 µm EVO C18 column by the proposed Excel-spreadsheet-based software, the effectiveness of the same optimal conditions into separation of solutes in different reversed-phase-type columns of the same size was tested. Indeed, a perfect resolution of the sample of interest is achieved in the chromatograms recorded in different columns under the optimal conditions determined for the Kinetex 5 µm EVO C18 column; see, as an example, Figure 5 for the application of optimal gradient conditions determined at eluent pH 3. Consequently, the optimal conditions derived for the separation of a sample on a certain column by the optimization algorithm could be successfully applied to other columns of the same type and size. Moreover, in Figure 5, the superiority of the core–shell technology columns and especially of the Kinetex EVO column is illustrated, since it is clear that a complete separation of test compounds was achieved within the minimum run time.

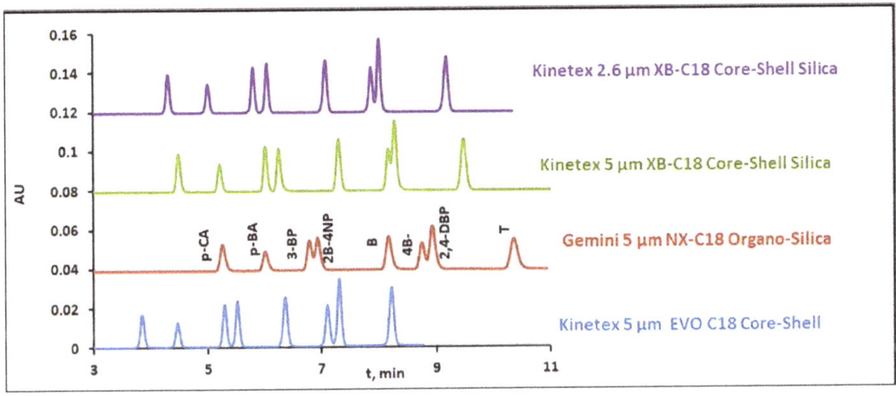

Figure 5. UV detected chromatograms of the mixture of 8 ionized and non-ionized solutes obtained on different columns under optimal gradient conditions found for eluent pH = 3 and Kinetex 5 µm EVO column. The elution order of solutes is shown in the Figure. See Table 1 for details of optimal gradient conditions.

The importance and the advantages of performing a computer-aided separation optimization are clearly shown in Figures 6 and 7. Figure 6 is a screenshot of the MS Excel spreadsheet 'isocr. optim.' used for isocratic separation optimization of test solutes in eluent pH 3 modified with MeOH. The retention data recorded in this eluent, depicted in Table S1, were analyzed by means of the Excel-spreadsheet optimization program following a procedure similar to that described above for separation optimization in eluent pH 3 modified with ACN. The optimal eluent, ϕ_{MeOH}= 0.575 was automatically found by pressing *Ctrl+w*. However, the selection of the optimal separation conditions is also possible from a good appreciation of the inset Graph A of Figure 6, which is also automatically created as described above. As shown in this figure, $t_R(max)$, decreases with increasing organic content ϕ in the mobile phase (purple circle markers), as is expected for a reversed-phase-type elution. However, the dependence of R_s (the resolution of the least resolved pair of adjacent solutes) on ϕ (depicted by the green diamond markers) is rather peculiar. For example, the resolution in mobile phases with either ϕ_{MeOH} = 0.46 or ϕ_{MeOH} = 0.52 is almost zero, which means that at least two solutes co-elute under the above isocratic conditions even though the run times, i.e., the values of $t_R(max)$ correspond to these eluent concentrations are longer than that in the optimal eluent with ϕ_{MeOH} = 0.575. Consequently, the foreknowledge of the precise dependence of R_s and $t_R(max)$ upon ϕ, provided by Excel-spreadsheet-based software adopted in this study and not by a trial-and-error method gives chromatographers a feel of confidence for the selection of the optimal conditions for a desired separation. Indeed, Figure 7 depicts a perfect resolution of the test solutes in the chromatogram recorded on the Kinetex 5 μm EVO C18 column and in the optimal eluent with ϕ_{MeOH} = 0.575. In contrast, the resolution of the same sample in a mobile phase with ϕ_{MeOH} = 0.5 is worse than that in an eluent ϕ_{MeOH} = 0.575 even though the separation time is longer, see also Figure 7.

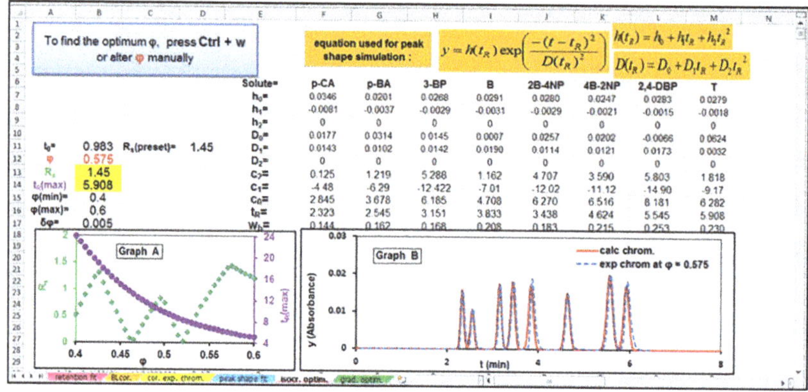

Figure 6. Screenshot of the MS Excel spreadsheet 'isocr. optim.' used for isocratic separation optimization of solutes in eluent pH 3 modified with MeOH. See the text for details.

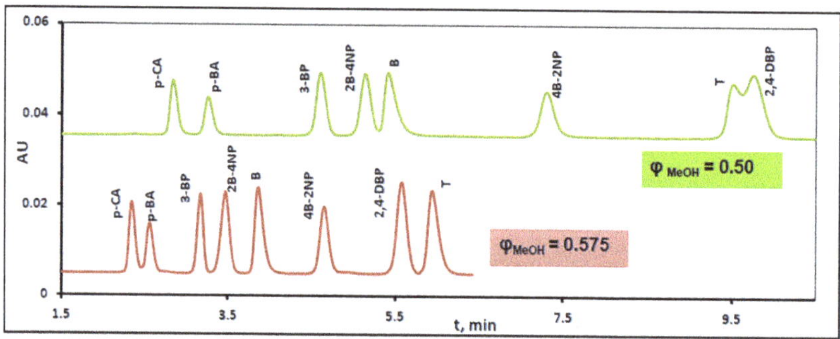

Figure 7. UV detected chromatograms of the mixture of eight ionized and non-ionized solutes obtained on Kinetex 5 μm EVO column under different isocratic conditions in eluent pH = 3 modified with MeOH. The elution order of solutes is shown in the Figure.

4. Conclusions

In this study, the whole separation optimization procedure of the test analytes under isocratic conditions as well as under single linear ϕ-gradients in different eluent pHs was successfully implemented by using an Excel-spreadsheet-based software, a user-friendly and widespread software platform, based on a few initial experiments for each eluent pH: three isocratic runs and two single linear ϕ-gradient runs performed in the studied range of mobile phase strength, ϕ. In the adopted optimization process, for computational simplicity, the solute retention parameters were obtained from the analysis of isocratic data, whereas a Gaussian function was used to fit peak shapes. The importance and the advantages of performing a computer-aided separation optimization compared with a trial-and-error optimization method were realized. The optimal separation conditions derived by the optimization algorithm for the separation of the sample of interest on a certain reversed-phase column were successfully applied to other same-type columns of the same size. The superiority of the core–shell technology columns and especially of the Kinetex EVO column was illustrated. In general, we consider that the results of this study could give chromatographers a feel of confidence for the selection of the optimal separation conditions for a sample of ionizable and neutral compounds in reversed-phase columns.

Supplementary Materials: The following are available online at www.mdpi.com/2297-8739/5/1/19/s1, Table S1: Experimental retention data, $t_R(exp)$ in min, of test solutes obtained on the Kinetex 5 µm EVO C18 column and under isocratic and linear gradient conditions in different eluent pHs modified with ACN or MeOH (PDF), Excel-spreadsheet-based software: "Optimization in eluent pH 3 modified with ACN" (ZIP).

Author Contributions: The experimental design was constructed and supervised by A.P.-L. The experiments were performed by A.M.M. and E.B. The data were analyzed by C.Z. The manuscript was drafted and written by A.P.-L. and C.Z.

Conflicts of Interest: The authors declare no conflict of interest.

References

1. Jupille, T.; Snyder, L.; Molnar, I. Optimizing multi-linear gradients in HPLC. *LC-GC Eur.* **2002**, *15*, 596–601.
2. Concha-Herrara, V.; Vivó-Truyols, G.; Torres-Lapasio, J.R.; García-Alvarez-Coque, M.C. Limits of multi-linear gradient optimization in reversed-phase liquid chromatography. *J. Chromatogr. A* **2005**, *1063*, 79–88. [CrossRef]
3. Fasoula, S.; Zisi, C.; Gika, H.; Pappa-Louisi, A.; Nikitas, P. Retention prediction and separation optimization under multilinear gradient elution in liquid chromatography with Microsoft Excel macros. *J. Chromatogr. A* **2015**, *1395*, 109–115. [CrossRef] [PubMed]
4. Debrus, B.; Lebrun, P.; Rozet, E.; Schofield, T.; Mbinze, J.K.; Marini, R.D.; Rudaz, S.; Boulanger, B.; Hubert, P. A new method for quality by design robust method optimization in liquid chromatography. *LC-GC Eur.* **2013**, *26*, 370–375.
5. Tyteca, E.; Liekens, A.; Clicq, D.; Fanigliulo, A.; Debrus, B.; Rudaz, S.; Guillarme, D.; Desmet, G. Predictive elution window shifting and stretching as a generic search strategy for automated method development for liquid chromatography. *Anal. Chem.* **2012**, *84*, 7823–7830. [CrossRef] [PubMed]
6. Fasoula, S.; Nikitas, P.; Pappa-Louisi, A. Teaching simulation and computer-aided separation optimization in liquid chromatography by means of illustrative Microsoft Excel spreadsheets. *J. Chem. Educ.* **2017**, *94*, 1167–1173. [CrossRef]
7. Canals, I.; Valkó, K.; Bosch, E.; Hill, A.P.; Rosés, M. Retention of ionisable compounds on HPLC. 8. Influence of mobile-phase pH change on the chromatographic retention of acids and bases during gradient elution. *Anal. Chem.* **2001**, *73*, 4937–4945. [CrossRef] [PubMed]
8. Andrés, A.; Rosés, M.; Bosch, E. Gradient retention prediction of acid–base analytes in reversed phase liquid chromatography: A simplified approach for acetonitrile–water mobile phases. *J. Chromatogr. A* **2014**, *1370*, 129–134. [CrossRef] [PubMed]
9. Christian, G.D.; Purdy, W.C. The residual current in orthophosphate medium. *J. Electroanal. Chem.* **1962**, *3*, 363–367. [CrossRef]
10. Horvath, C.; Melander, W.; Molnar, I. Liquid chromatography of ionogenic substances with nonpolar stationary Phases. *Anal. Chem.* **1977**, *49*, 142–154. [CrossRef]
11. Wiczling, P.; Kaliszan, R. pH gradient as a tool for the separation of ionizable analytes in reversed-phase high-performance chromatography. *Anal. Chem.* **2010**, *82*, 3692–3698. [CrossRef] [PubMed]
12. Nikitas, P.; Pappa-Louisi, A.; Zisi, C. pH-gradient reversed-phase liquid Chromatography of ionogenic analytes revisited. *Anal. Chem.* **2012**, *84*, 6611–6618. [CrossRef] [PubMed]
13. Fasoula, S.; Zisi, C.; Nikitas, P.; Pappa-Louisi, A. Retention prediction and separation optimization of ionizable analytes in reversed-phase liquid chromatography by organic modifier gradients in different eluent pHs. *J. Chromatogr. A* **2013**, *1305*, 131–138. [CrossRef] [PubMed]
14. Schoenmakers, P.J.; Billiet, H.A.H.; de Galan, L. Description of solute retention over the full range of mobile phase compositions in reversed-phase liquid chromatography. *J. Chromatogr. A* **1983**, *282*, 107–121. [CrossRef]

MDPI
St. Alban-Anlage 66
4052 Basel
Switzerland
Tel. +41 61 683 77 34
Fax +41 61 302 89 18
www.mdpi.com

Separations Editorial Office
E-mail: separations@mdpi.com
www.mdpi.com/journal/separations

www.ingramcontent.com/pod-product-compliance
Lightning Source LLC
LaVergne TN
LVHW070627170726
843515LV00004B/198